Laser Surface Treatments for Tribological Applications

Edited by

Jeyaprakash Natarajan., Ph.D.,

Research Assistant Professor
Additive Manufacturing Center for Mass Customization
Production
Graduate Institute of Manufacturing Technology
National Taipei University of Technology, Taipei-10608
Taiwan

&

Prof. Che-Hua Yang., Ph.D.,

Director and Professor
Additive Manufacturing Center for Mass Customization
Production
Graduate Institute of Manufacturing Technology
National Taipei University of Technology, Taipei-10608
Taiwan

Laser Surface Treatments for Tribological Applications

Editors: Jeyaprakash Natarajan and Che-Hua Yang

ISBN (Online): 978-981-5036-30-5

ISBN (Print): 978-981-5036-31-2

ISBN (Paperback): 978-981-5036-32-9

need for a court order if at any point you breach any terms of this License Agreement. In no event will any delay or failure by Bentham Science Publishers in enforcing your compliance with this License Agreement constitute a waiver of any of its rights.

3. You acknowledge that you have read this License Agreement, and agree to be bound by its terms and conditions. To the extent that any other terms and conditions presented on any website of Bentham Science Publishers conflict with, or are inconsistent with, the terms and conditions set out in this License Agreement, you acknowledge that the terms and conditions set out in this License Agreement shall prevail.

Bentham Science Publishers Pte. Ltd.
80 Robinson Road #02-00
Singapore 068898
Singapore
Email: subscriptions@benthamscience.net

BENTHAM SCIENCE

CONTENTS

FOREWORD

When Dr. Jeyaprakash and Prof. Che-Hua Yang invited me to write this preface, I was honored and delighted to take this opportunity to be familiarized with this exceptional work.

LASERS are being used for almost everything today, but they took a long time to seem useful. Albert Einstein invented laser technology in the early 1900s. Until 1960, there were no developments, and then the first laser was built at Hughes Research Laboratories. It is exciting to know the laser journey that Albert Einstein laid the foundations of in 1917 by predicting the phenomenon "Stimulated Emission." Today it has become a fundamental phenomenon for the operation of all kinds of lasers. Valentin Fabrikant hypothesized the use of stimulated emission to amplify radiation . As a similar laser, its development was also speedy.

Further, Charles Townes, Nikolay Basov, and Alexander Prokhorov developed the theorem of stimulated emission of microwaves in 1950, for which they received a noble prize in Physics. In 1959, Gordon Gould proposed that the stimulated emission can be used to amplify light. His extensive research described that the optical resonator could create a narrow beam of light called "Light Amplification by Stimulated Emission of Radiation." Right after his invention, in 1960, Theodore Maiman built the first working laser at Hughes Research Lab in Malibu. The laser, which used synthetic ruby as its active medium, emitted a bright red beam of light with a wavelength of 694.3 nm. The first application by this ruby laser was for military range finders. Surprisingly still, this technology is used commercially for drilling holes in diamonds due to its high power. In 1963, Kumar Patel developed the Carbon Dioxide (CO_2) laser at AT&T Bell labs. It is economical and has high efficiency than the ruby laser. Therefore, it became the leading laser technology for many industries and is being continuously used for more than 50 years.

In the 1960s, the CO_2 laser became famous for materials processing applications. In 1967, Peter Houldcroft of TWI in Cambridge utilised an oxygen-assisted CO_2 laser beam to cut a 1 mm thick steel sheet, which was the first commercial application of Laser Materials Processing.These applications grew in scope, leading to the development of a small and inexpensive CO_2 slab laser in the 1980s, ushering in a new "Laser Materials Processing" era. Since then, the laser has played a significant role in the materials industry for various applications, such as metal cutting, welding, drilling, and organic processing materials, such as rubber, foam, and plastic. In 2009, the largest and highest-energy laser in the world was developed in the National Ignition Facility at Lawrence Livermore National Laboratory. In the same year, NASA launched the Lunar Reconnaissance Orbiter. It will use a laser to gather data about the high and low points on the moon. This information may create 3D maps to identify lunar ice locations and safe landing sites for future spacecraft. In 2019, MIT researchers investigated a 1.9 micrometers wavelength thulium laser to excite water molecules near a microphone, which transmit an audible signal. This signal was as loud as a regular conversation. This technique can send secret messages for military applications and advertising. Currently, the laser is being used in a variety of industries, including medicine, automobiles, aircraft, and others. Finally, very slowly but surely, "the laser is living up to its billing."

Prof. Jia-Chang Wang
Department of Mechanical Engineering
National Taipei University of Technology
Taipei-10608
Taiwan

PREFACE

Metals for industrial applications require several properties, such as ductility, malleability, hardness, strength, corrosion resistance, thermal expansion, availability, reusability, *etc*. When it comes to tribological application, hardness, strength, and surface properties are the primary necessities. However, achieving all the properties in a single metal is not easy. The product developers have to pick the appropriate metal as per the application requirements by understanding the wide variety of metals and their functional properties. The metallic components are assembled to deliver the relative motion, however, the friction will be generated due to the interaction between the metallic surfaces, and this interaction will lead to 'wear' of the metallic components. Wear in mechanical components reduces the plant efficiency because of power losses, and in rare cases, it is catastrophic. Wear is a vital cause of malfunctioning a mechanical system, and it is a serious issue that needs to be addressed in tribological applications.

Lasers are used in many industries, and their applications in various fields are only growing with time. The laser-assisted machinery highlights how lasers have helped us to be at the forefront of technology, making rapid changes possible, including the improvement of tribological properties which is a battle that many scientists throughout the world are battling to limit the wear losses from friction between moving parts. The use of lasers has become a significant source and splendid tool for various surface modification methods, such as hardening, melting, alloying, cladding and texturing, *etc*. Laser surface treatments offer extensive promises to accomplish preferred surface and tribological properties.

These surface properties have been studied with a wide range of characterization methods, such as the chemical composition of the surface being analyzed with the energy dispersive, Auger electron, glow discharge, optical spectroscopy and X-ray spectroscopy. The microstructure and morphology of the surface are studied by using light optical, scanning electron and transmission electron microscopy. The tribological characterization includes the evaluation of surface roughness, friction, and wear aspects of the surfaces from macro to nano scale. The surface profile can be assessed by using contact and non-contact type surface profilometers.

The challenges in laser surface treatments and tribological issues are both difficult and interesting. People are working with enthusiasm, tenacity, and dedication to develop new techniques/methods of analysis and provide excellent solutions. In this new age of global interconnectivity and interdependence, it is necessary to provide the latest details for both professionals and students, with state-of-the-art knowledge on the frontiers in laser surface treatments for tribological applications. This book is a good step in that direction.

Jeyaprakash Natarajan and Che-Hua Yang
Additive Manufacturing Center for Mass Customization Production
Graduate Institute of Manufacturing Technology
National Taipei University of Technology, Taipei-10608
Taiwan

ACKNOWLEDGMENTS

The editor and co-editor would like to thank National Taipei University of Technology (NTUT), Taipei-10608, Taiwan for the valuable support and extended facility to carry out this book project in successful manner. In addition to that editor(s) thank to Mrs. Humaira Hashmi, Editorial Manager, Bentham Science Publication for his valuable guidance and impressive official help on this book. The enthusiasm and thoughtfulness of NTUT faculty, students, staff, and community partners over the years inspired us to create this resource for other academic colleagues.

DEDICATION

To the many teachers across five countries I learnt so much from, and my loving family who unconditionally supported me throughout my busy working life.

List of Contributors

D. Gunasekar	Department of Mechanical Engineering, Jayaram College of Engineering and Technology, Trichy- 621014, Tamil Nadu, India
D. Raj Kumar	Department of Mechanical Engineering, MAM School of Engineering, Trichy - 620026, Tamil Nadu, India
G. Vignesh	Karpagam Academy of Higher Education, Coimbatore-641 021, India
K. Ganesa Balamurugan	Department of Mechanical Engineering, IFET College of Engineering, Villupuram- 605108, Tamil Nadu, India
K. Tejonadha Babu	Department of Mechanical Engineering, Kallam Haranadhareddy Institute of Technology, Guntur-522019, Andhra Pradesh, India
Kaushik N. Ch	Department of Mechanical Engineering, School of Engineering and Technology, BML Munjal University, Gurgaon- 122413, Haryana, India
Krishna Kishore Mugada	Department of Mechanical Engineering, Indian Institute of Technology Delhi, New Delhi- 110016, India
Mahendra Babu Kantipudi	Department of Vehicle Engineering, National Taipei University of Technology, Taipei, Taiwan Department of Mechanical Engineering, Vishnu Institute of Technology, Bhimavaram, Andhra Pradesh, India
Milon Selvam Dennison	School of Engineering and Applied Sciences, Kampala International University, Western Campus, Uganda
Muralimohan Cheepu	Super-TIG Welding Co., Limited, Busan- 46722, Republic of Korea
Nandhini Ravi	Department of Metallurgical and Materials Engineering, National Institute of Technology, Tiruchirappalli-620015, Tamil Nadu, India
R. Dinesh Kumar	Department of Mechanical Engineering, Indian Institute of Technology, Guwahati-781039, Assam, India
S. Venukumar	Department of Mechanical Engineering, Vardhaman College of Engineering, Hyderabad-501218, Telangana, India
T. Prabakaran	Department of Mechanical Engineering, MIET Engineering College, Trichy - 620007, Tamil Nadu, India
T. Vijaya Babu	Department of Mechanical Engineering, Vardhaman College of Engineering, Hyderabad-501218, Telangana, India
Varthini Rajagopal	Department of Mechanical Engineering, Government College of Engineering Srirangam-620012, Tamil Nadu, India
Venkata Charan Kantumuchu	American Society for Quality, Oklahoma City, Oklahoma, the United States/Mauser Packaging Solutions, Memphis, Tennessee, USA
Venkateswarlu Devuri	Department of Mechanical Engineering, Marri Laxman Reddy Institute of Technology and Management, Hyderabad, India
Yaojung Shiao	Department of Vehicle Engineering, National Taipei University of Technology, Taiwan

CHAPTER 1

Metals and their Tribological Applications

Mahendra Babu Kantipudi[1,2,*] and **Yaojung Shiao**[1]

[1] *Department of Vehicle Engineering, National Taipei University of Technology, Taiwan*

[2] *Department of Mechanical Engineering, Vishnu Institute of Technology, Bhimavaram, Andhra Pradesh, India*

Abstract: The selection of metals is an essential and tricky step to achieve the product's best outcomes. Metals for industrial applications require several properties, such as ductility, malleability, hardness, strength, corrosion resistance, thermal expansion, availability, reusability, *etc*. When it comes to tribological applications, hardness, strength, and surface properties are the primary necessities. Alloying, heat treatment and surface treatment are the various techniques to attain these metals' properties. Due to extensive research for a long time, many metal alloys already exist for tribological applications. However, achieving all the properties in a single metal is not possible. The product developers have to pick the appropriate metal according to the application requirements by understanding the wide variety of metals and their functional properties. Hence, this chapter gives a comprehensive reference for the various metal alloys and their applications. Firstly, metals are broadly classified based on their primary composition. Then, the metallurgical characteristics, alloying elements, physical and mechanical properties of the various metals are explained. Lastly, the tribological applications of those metals are discussed.

Keywords: AISI 304 Stainless steel, Aluminium alloys, Babbitt, Bearing materials, Cast iron, Copper alloys, Ferrous metals, Friction, Hardness, Mechanical properties, Metallurgy, Metal alloys, Nodular cast iron, Non-ferrous metals, Steels, Superalloys, Tool steels, Tribological applications, Wear.

INTRODUCTION

From the ancient days, metals are playing a vital role in human civilization and engineering developments. They are reliable, strong, and smooth elements that have good conductivity of heat and electricity. Metals are characteristically ductile and malleable due to their nature of metallic bonding. Each of the metal atoms gives its valence electrons to establish an electron enclosed around the positively charged ions. This band of several atoms forms a bond and becomes a

---
* **Correspondence author Mahendra Babu Kantipudi:** Department of Vehicle Engineering, National Taipei University of Technology, Taiwan, Email: mahendra.k4u@gmail.com

Jeyaprakash Natarajan and Che-Hua Yang (Eds.)

solid structure, as shown in Fig. (**1**). At applied shear force, metal ions can slide over each other and rebuild their bond without losing their mutual electron bonding.

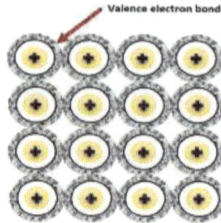

Fig. (1). Mechanism of metallic bonding.

The crystal structure is an essential criterion to understand metal behavior. When a liquid state material cools, it forms a solid with some pattern. This solid formation pattern is called a crystal structure. It consists of atoms in a uniform, repeating, and three-dimensional (3D) order. The smallest repeating 3D arrangement is called a unit cell. Most of the metal structures have existed in three crystalline patterns. They are namely face-centered cubic (FCC), body-centered cubic (BCC), and hexagonal close-packed (HCP).

Fig. (**2**) shows the placement of the atoms in the three types of patterns. The packing density and some of the properties of the metals depend on these metal structures. FCC has a high packing value (ratio of the volume of atoms and the unit cell), *i.e.*, 0.74, whereas these atomic packing values for BCC and HCC are 0.68 and 0.74, respectively. Chromium, vanadium, α-iron, and tungsten are some of the metals with a BCC structure, whereas aluminum, copper, gold, lead, nickel, and silver are the FCC structured metals. Besides, metals like beryllium, magnesium, titanium, and zinc are the HCP structure. Few metals can give more than one crystal structure; this ability is known as polymorphism. Coming to fundamental solids, this facility is often called allotropy. The crystal structure of pure iron is BCC at room temperature and FCC at above 900 ^{0}C. Polymorphic transformation changes the properties and density of the metal.

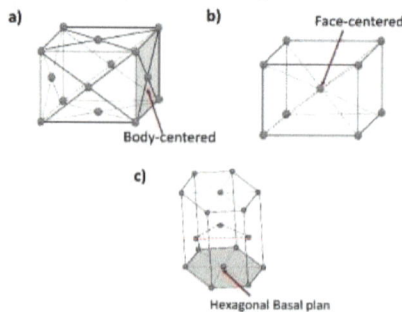

Fig. (2). Atoms arrangement in (**a**) BCC, (**b**) FCC, and (**c**) HCP systems.

In most engineering applications, specifically in tribological applications, metals have to endure mechanical and thermal stresses. Hardness, toughness, tensile strength, surface friction coefficient, thermal conductivity, and thermal expansion are the crucial metal properties for tribological applications. Hardness is a degree of resistance to localized deformation due to either mechanical indentation or scratch. Toughness is the capability of a metal to absorb energy and deform plastically without fracturing. The metal's tensile strength is the maximum stress that it can withstand while being stretched before breaking. The friction coefficient is a proportionality constant, which indicates that the tangential resistive force varies linearly with the normal load. Thermal conductivity is the degree of the metal's ability to conduct heat. Lastly, thermal expansion is the inclination of metal to change its shape, dimensions, and density in reaction to a temperature change.

FERROUS METALS

The Latin word 'ferrous' refers to a metal that contains iron. The base metal is iron, and the ferrous metal's crucial element is a carbon (C). Other metal compositions are added to achieve the essential properties. The amount of carbon in the ferrous alloys affects the metal properties. Ferrous metals have good magnetic and mechanical properties. Ferrous metals are the most convenient metal than any other metal because of the reasons like an abundance of raw materials, good mechanical properties, ease of extraction and processing, and ease of formation. However, corrosion is a major problem for these metals.

Classification of Ferrous Metals

Fig. (**3**) shows that ferrous metals can be broadly classified into steels and cast irons depending on the carbon and other alloys compositions. The iron-carbon diagram is often used to comprehend the various phases of carbon steel and cast iron. This diagram's X-axis shows the percentage of carbon that starts at 0% and ends at 6.67%. Up to 0.008% carbon content, the metal is pure iron, which has poor mechanical properties.

As shown in Fig. (**4**), ferrous metals can be classified with carbon percentage. Ferrous alloys with a carbon composition of 0.008% to 2.14% are called steels. They are ductile and strong enough. If the carbon composition increases by more than 2.14%, then the alloy reaches the cast iron phase. Cast iron is tough; nevertheless, its brittleness strictly bounds its applications and methods for manufacturing.

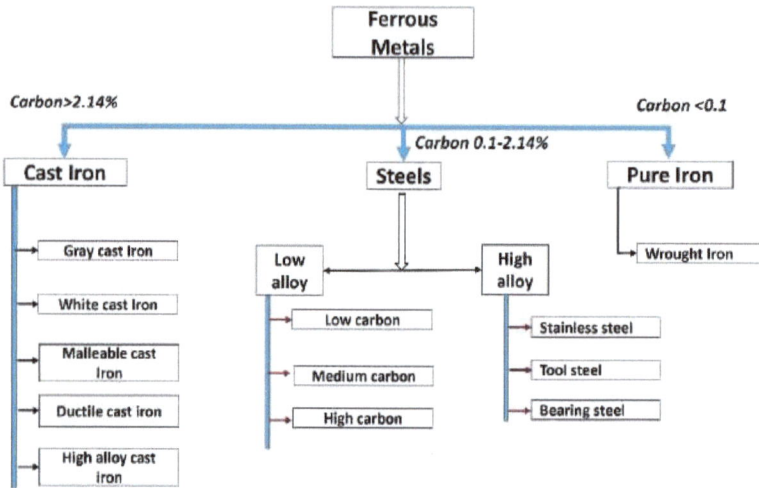

Fig. (3). Classification of ferrous metals.

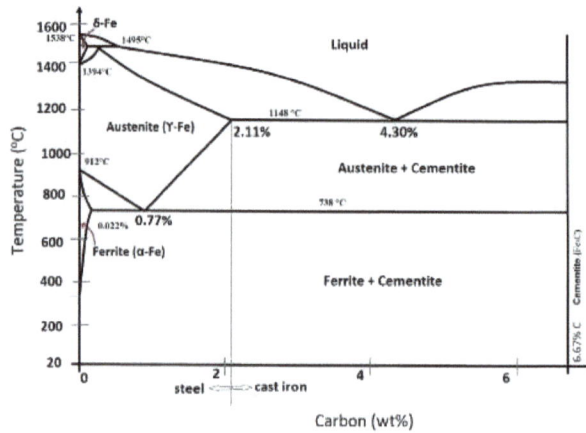

Fig. (4). Iron-Carbon phase diagram.

CAST IRON

Cast iron is the oldest and economical metal for engineering applications. It contains a higher carbon percentage than steel; therefore, it can benefit from eutectic solidification. Table **1** contains the carbon and other alloys percentages in various types of cast irons. The eutectic point happens in the iron-carbon phase diagram at a temperature of 1148 °C, with a carbon composition of 4.26%. Cast iron usually comes under ferrous alloy, which contains more than 2% carbon and more than 1% silicon. Cast iron metal has flowability at the molten phase due to the higher carbon and silicone compositions. Therefore, they are perfect for casting activities. It is an excellent metal for tribological applications due to its high hardness, wear resistance, and high-temperature stability properties. The mechanical and physical properties of various cast iron metals are listed in Table **2**.

Table 1. Compositions of alloys in different cast iron metals.

-	Carbon (%)	Manganese (%)	Silicon (%)	Phosphorus (%)	Sulfur (%)
Gray cast iron	2.5-4.0	0.25-1.0	2.1-2.3	0.05-0.1	0.02-0.25
White cast iron	1.8-3.6	0.25-0.8	0.5-1.9	0.06-0.2	0.06-0.2
Malleable cast iron	2.0-2.9	0.2-1.2	1.1-1.6	0.01-0.18	0.04-0.18
Ductile cast iron	3.0-4.0	0.1-1.0	1.8-2.8	0.01-0.1	0.01-0.03

Table 2. Mechanical and physical properties of various cast iron metals.

-	Gray cast iron	White cast iron	Malleable cast iron	Ductile cast iron
Ultimate tensile strength (MPa)	207 Mpa	345-689	207-621	345-650
Hardness (HB or HV)	174-210	350-807	149-321	150-352
Wear resistance	Moderate	High	Moderate	Moderate
Density (kg/m^3)*10^3	6.8-7.15	7.59-7.77	7.2-7.45	6.64-7.42
Elastic modulus	Low	High	High	High
Thermal expansion (1/^{0}K)	5.8 E-06	6.5 E-06	6.7 E-06	8.2 E-06

Iron carbide, pearlite, and austenite are the probable microstructures for cast iron. These structures can be achieved with different cooling rates. Pearlite can be achieved at slow cooling. At rapid cooling, carbon forms carbides instead of graphitizing. Cast irons can be classified into different metals depending on the above microstructures. They are gray cast iron, white or chilled cast iron, ductile (or nodular) cast iron, malleable cast iron, and compacted graphite cast iron [1].

Gray Cast Iron

Gray cast iron is one of the prevalent forms of cast iron, in which carbon subsists as graphite flakes, and it is an alloy of iron-carbon-silicon. It can be achieved by the addition of greater than 1% of silicon. Silicon stimulates the development of flake graphites. Graphite flakes exist when the iron carbide (cementite, Fe_3C) detaches into alpha graphite and (α) ferrite. The graphite establishment is controlled with the silicon percentage and the slow cooling rates in the solidification progression. The fractured surface of this cast iron is grayish due to the existence of these graphite flakes. The carbon percentage in the existing cast irons varies between 2.1 and 4.5%. It has properties like high compressive strength, good damping properties, and high resistance to wear. However, due to the graphite flakes, gray cast irons are fragile and delicate in tension.

Pearlite and ferrite are the two possible microstructures of the gray iron that depend on the casting's cooling speed. It also has good tribological properties like less friction and good wear resistance. It is common for the piston ring and cylinder liner interaction in the IC engine [5]. The flake graphite offers a good lubrication film, which provides brilliant wear and friction features in a dry sliding contact [6]. An extensive range of applications is there for gray cast irons. Internal combustion (IC) engine components, brake drums, machinery housings, hydraulic pumps, compressors, pipefittings, and pumps are a few examples.

White Cast Iron

White cast iron is one more regular cast iron. It can be achieved by heat treatment of gray cast iron with rapid quenching. With the combination of low Si content (0.5 to 1.5%) and faster cooling rates, cementite cannot get decomposed. Therefore, it retains as brittle cementite. White cast iron has 1.8% -3.6% of carbon, 0.5% -1.9% of silicon, and 1% – 2% of manganese. In this metal, carbon exists as iron carbides. The fracture face of white cast iron looks white due to a large fraction of carbides. It has outstanding resistance against wear and abrasion due to the huge masses of carbides. However, they are extremely difficult for machining operations. Therefore, their usage is limited to wear-resistant applications. It is used for shot blasting nozzles, ball mill liners, crushers, and as rollers of rolling mills and rings in pulverizers.

Malleable Cast Iron

Malleable cast iron is a soft and ductile metal. It can be achieved by the heat treatment of white cast iron. White cast iron is firstly heated to around 920^{0}C temperature, and then it is left to prolonged cooling. During this slow cooling, graphite separates at a much slower rate. It has enough time to form spheroidal elements rather than flakes. This metal is stronger, with a significant amount of ductility. This metal is used for railways, connecting rods, naval, and other heavy-duty applications.

Nodular (Ductile) Cast Iron

A small amount of magnesium or cerium agent to the gray cast iron gives a noticeably different microstructure, as shown in Fig. (**5**). In this process, graphite is formed into nodules or sphere-like elements.

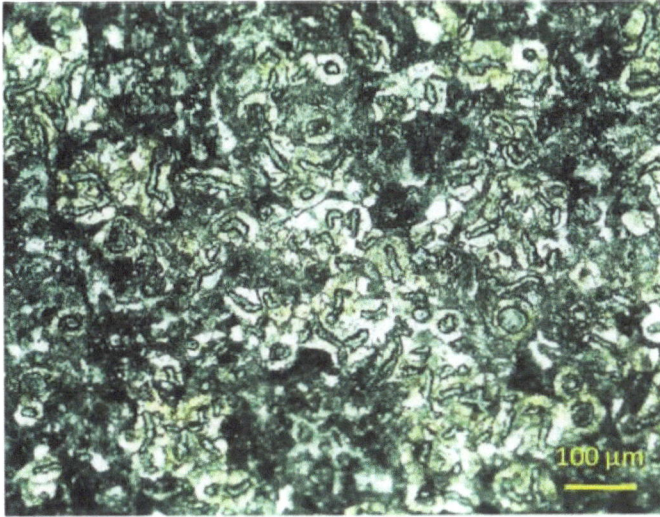

Fig. (5). Microstructure of nodular cast iron.

The matter surrounding these nodule elements is either ferrite or pearlite, as per the heat treatment procedure. Nodular cast iron is a stronger ductile metal than gray cast iron. It is used for pump structures, shafts, and locomotive components. It is not the best choice for heavy wear industrial applications due to its limited wear resistance. However, several techniques like laser surface alloying are used to enhance the tribological properties of nodular cast iron [7].

High-Alloy Cast Irons

The tribological requirements like better strength, high wear resistance, corrosion resistance, and stability at elevated temperatures can be achieved by adding smaller amounts of alloy elements like chromium, nickel, or molybdenum. One high alloy cast iron with exceptional abrasion resistance property is high chromium white cast irons (HCWCI). All HCWCI are hypoeutectic alloys with 10 - 30% of chromium and 2 - 3.5% of carbon [8]. These metals are excellently suitable for coal grinding parts, mining equipment, milling tools, and slurry pumps. Similarly, nickel-alloyed (13 to 36% Ni) cast irons and the high-silicon (14.5% Si) gray irons are used in applications requiring corrosion resistance. Moreover, nickel-alloyed gray and ductile irons are useful for high-temperature service.

Graphite, cementite, austenite, ferrite, and pearlite are the possible physical structures of cast iron that influence its properties. Annealing, quenching, tempering, and surface hardening are the heat treatment concepts that give these various microstructures. Surface treatment is an essential technique for tribology applications to achieve maximum wear resistance without losing the structure's toughness. Laser, flame, and induction can be used as heating sources for this hardening. Up to 600 Vickers of hardness can be achieved by these operations. The depth of the surface hardening is generally about 1.8 mm. Nitriding is one surface hardening method to achieve surface hardness up to 900 Vickers. This is useful for distinct alloy cast irons having aluminum and chromium. Cyaniding or the salt-bath method is used to nitriding the gray and nodular cast irons. Cyaniding not only enhances wear resistance but also increases fatigue strength and corrosion resistance. Advanced surface treatments like laser alloying using WC-12%Co and Cr3C2-25%NiCr alloy powders improve the wear resistance of the nodular iron surface [9].

Surface properties are also very extremely important for the tribological application to reduce friction. Electro discharge machining, photochemical milling [10], electrochemical machining [11], laser beam machining [12], and magnetorheological fluid-based finishing process [13, 14] are the advanced surface texturing processes.

STEELS

Undoubtedly, steels are the best and most useful metal for tribological applications. They have extensive and diversified usage. For instance, AISI 52100 steel is applicable for the ball and roller bearings after a special heat treatment to reduce austenite content and guarantee dimensional firmness. AISI440C steel is applicable for roller contact bearings due to its corrosion resistance and high-temperature properties. Manganese alloy steel can be used for high-impact resistance and wear resistance applications like mining operations and railways. Steel surfaces are being treated with wear-resistant material coatings to enhance their tribological properties. Various steels that are applicable in tribological applications are explained in the next sections.

As already mentioned, steels are fundamentally alloys of iron and carbon. The interstitial sites of the Fe structure are occupied by carbon. Alloying elements, such as nickel, chromium, manganese, silicon, sulfur, molybdenum, vanadium, and tungsten are mixed to enhance the alloy's physical properties and improve the corrosion resistance. Steels can be majorly classified into low alloy carbon steels and high alloy steels depending on their carbon composition.

Low Alloy Carbon Steels

It contains less than 1% of carbon and little manganese, sulfur, silicon, and phosphorus. Even though alloying and residual elements affect this type of steel's characteristics, it majorly depends on the carbon content percentage. Carbon steels are low-cost and perfect for large components. In the tribological application, the primary requirement is wear resistance. Steels are undoubtedly the most useful metals in this context. Carbon steels are divided into four subgroups as follows.

Low Carbon Steel

Carbon content in low carbon steel is typically 0.04% to 0.30%. It covers an excessive variety of forms, from sheet metals to structural elements. Other alloy elements are added to achieve wanted properties. For instance: maintaining the low-level carbon and adding aluminum to achieve drawing excellence and maintaining high carbon manganese content levels to achieve structural stability. AISI 1008, 1010, 1015, 1018, 1020, 1022, 1025 are the popular low carbon steels.

Medium Carbon Steel

The carbon content in medium carbon steel is around 0.30 to 0.45%. This steel's hardness and tensile strength are higher than low carbon steel due to the increased carbon content. However, the ductility, machinability, and weldability of this metal are lower than low carbon steel due to higher carbon content. AISI 1030, 1040, 1050, 1060 are the general medium carbon steels.

High Carbon Steel

The carbon composition in high carbon steel is around 0.45 to 1.5 percent. This steel is very hard to cut, bend, and weld due to the high carbon percent. Heating is required to regulate the mechanical and physical characteristics of steel after welding. This type of carbon steel is used for hard applications like cutting tools, wear-resistant steel, *etc*. AISI 1080, 1095 are the few high carbon steels.

Carbon steel exists in an austenite state at 750 to 1000° C temperature. This austenite is restructured into carbides and ferrite when it cools to room temperature. The rate of cooling is very important in this reconstruction. The choice of the accurate quench rate is directed by the alloy's temperature-time transformation (TTT) characteristics. Fig. (**6**) shows the effect of cooling rate and initial temperatures on steel microstructure.

Pearlite Steel

This steel contains a structure of alternative layers of ferrite and carbides. This structure attains by heating the steel to above 1500°C temperature, then it is held for a while to dissolve all carbide, and cooled slowly to room temperature. This steel has high wear resistance and good strength. A larger amount of pearlite in steel increases the wear resistance. Increased carbon content increases the carbides and pearlite section in the arrangement, consequently increasing the hardness and wear resistance. Moreover, the hardness of the metal and pearlite grain size are controlled by quenching rates.

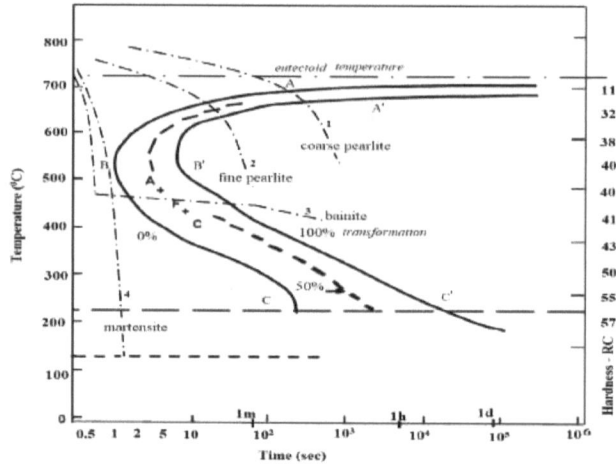

Fig. (6). TTT curve showing the transformations.

Martensitic Steel

It is a kind of hardened crystalline structured steel that is formed over distribution-less transformation process. It is formed in carbon steels with a rapid cooling (quenching) of the austenite by preventing the atoms from diffusing out of the crystalline structures. It is hard and a more wear resistant steel than other carbon steels. However, it is comparatively brittle until tempered by reheating after quenching. Tempering is one method to remove internal stresses in martensitic steel. Doing so increases its resistance to shock, making it less likely to break upon impact. Tempered martensitic steel maintains the balance of good strength and wear resistance.

Bainite

It is formed by being cooled faster than pearlite but slower than martensite. Bainite has plate-shaped microstructures, whereas martensite has a lengthy oval-shaped structure. Bainite is regularly chosen because it does not need tempering after being hardened. This structured steel has the same wear resistance as martensitic steels but with greater toughness.

Austenite and Ferrite

If enough amount of manganese is added, the carbon steels can stabilize austenite at room temperature. Austenitic steels have better wear resistance compared to ferritic steels at the same carbon content. It is high-impact resistant steel used for mining and soil-moving machinery.

It is clear from the above discussion that the wear resistance, hardness, and toughness of the carbon steels can be controlled by carbon percentage and the cooling rates. However, alloying of other elements are required to achieve other properties like corrosion resistance, stability at higher temperatures, a combination of hardness and toughness, *etc*. Table **3** listed various alloying elements and their effect on steel.

Table 3. Effect of alloying elements.

Name of the alloy	Achieved properties
Carbon	Strength and hardness
Manganese	Hardness, Corrosion resistance
Silicon	Sound metal, deoxidizer, impact resistance
Nickel	Strength and hardness without affecting ductility of the material
Chromium	Hardness, strength, wear resistance, and corrosion resistance
Molybdenum	has smaller grain size. It improves the elastic limit, wear resistance, and fatigue strength.
Titanium	Improve toughness and corrosion resistance
Vanadium	Increases hardness
Copper	Corrosion resistance, improve machinability and formability
Cobalt	The element for martensitic steel increases hardness

Stainless Steels

Steels become high corrosion-resistant metals by adding distinct alloying elements, particularly a minimum of 10.5% Cr alongside Ni and Mo. The name derives from its great corrosion resistance, such that they are stain-less.

Stainless steels are mainly of three types based on their microstructure, namely austenitic, ferritic, and martensitic. Austenitic steels have the highest corrosion resistance. Ferritic and austenitic steels are not heat treatable. They are hardened and strengthened by cold working. At the same time, martensitic steels are heat treatable. AISI has recognized a three-digit classification for stainless steel. Table **4** shows the composition details of various series of stainless steel.

Austenitic stainless steel is useful for a large number of applications due to its excellent corrosion resistance. It has the best combination of carbon, chromium, and nickel elements to achieve corrosion resistance. Therefore, it can be used under many chemicals, high temperatures, and aggressively corrosive conditions. 200 series and 300 series steel come under this class. 300 series stainless steel (AISI 304, *etc.*) are mostly used in tribological applications. Fig. (7) shows the microstructure of the AISI 304. Advanced coating techniques can improve the wear resistance of this alloy steel metal [15].

Table 4. Composition details of various series of stainless steels.

AISI grade	Type/Major alloys	C%	Cr %	Ni %	Mn %	Mo %	Other (S, P, and Si) %
200 series	Austenitic steels/chromium-nickel-manganese	0.15	14-19	3.50-6.00	4-15.5	0-1.5	0.5-1.25
300 series	Austenitic steels/composition of chromium-nickel	0.02-0.15	17-20	8-12	0-2	2	0-1.5
400 series	Martensitic or ferritic steels /chromium	0.1-1	12-29	1-2	1	0-1.2	0-1
500 series	Martensitic steels/low chromium	0.1	4-6	-	1	0.4-065	0.5-1.5
600 series	Martensitic/chromium-vanadium steel	0.5	15-18	3-7.5	1	-	0.5-5 (Cu, Al, or Nb)

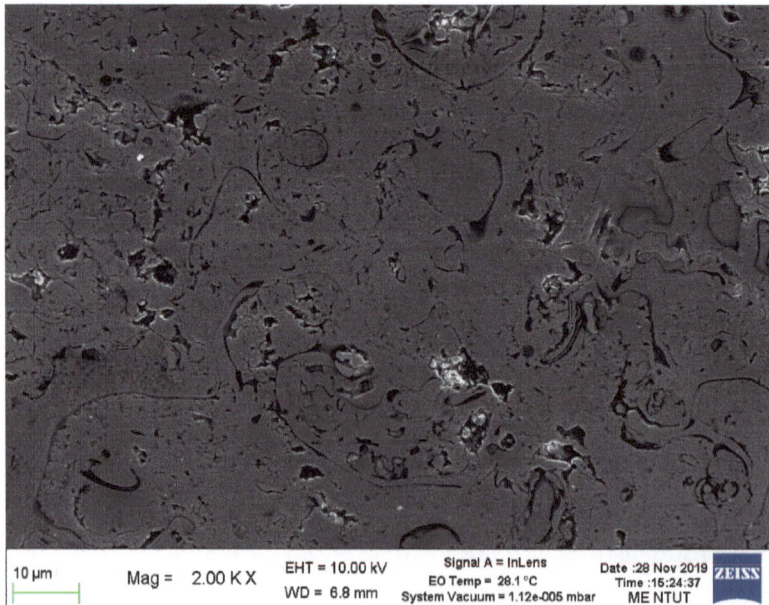

Fig. (7). Microstructure of AISI 304 stainless steel.

Martensitic stainless steel has more than 11.5% chromium, and little nickel content. This type of steel can be hardened by heat treatment. Therefore, the required properties can be obtained. However, the corrosion resistance of this steel is lower than austenitic stainless steel. The 400 series steel comes under this category. The mechanical and physical properties of several popular stainless steels are listed in Table **5**.

Table 5. Mechanical and physical properties of popular stainless steels [16].

AISI/ASTM Grade	Density (kg/m^3)	Hardness (HB or HV)	Tensile strength (MPa)	Modulus of elasticity (GPa)	Thermal expansion $(1/^0K) \times 10^{-06}$
201	7830	241	750-950	197	17.5
202	7800	241	680-880	197	18.4
304	7900	150	500-720	197	17.4
304HN	7900	200	530-740	200	19.4
305	7900	183	500-650	193	17.4
316	8030	150	530-680	196	17.4
316H	8030	217	520	198	16.8
321	7900	160	500-700	198	17.5
347	7890	160	520-720	198	18.7
310 MoLN	8000	217	540-740	195	17.0
409	7610	183	380-560	220	12.0
410	7640	210	400-600	220	11.7
420	7680	225	600-700	215	12.0
501	7750	321	400-680	200	11.2
630(17-4)	7750	400	1070-1379	196	11.4
631	7800	441	1030-1275	200	11.4
15-5PH	7750	420	1020-1379	196	11.2

Manganese Steel

Manganese steel, also called manga alloy, contains 12-14% manganese and 1.15% carbon. They are used for high toughness and wear-resistance necessary applications like mining equipment, ore handling equipment, and earthmoving machinery. Austenite structure can be stabilized in high carbon steels by adding a high amount of Mn alloy content.

Other elements like molybdenum and silicon are added to include carbides in an austenitic matrix so that it can be heat treatable or hardenable steel. Manganese steels also have considerable capacity for work hardening. During heavy working operations, manganese steel transforms to martensite. Therefore the surface of these steels becomes harder and wear-resistant.

One of the famous conventional manganese steels is Hadfield steel. It has 11 to 14 percent manganese and 1.1 to 1.25 percent carbon. It is best appropriate for heavy impact and scoring abrasion. One more well-known manganese austenitic steel is the lean manganese high carbon steel. It is not as strong as Hadfield steel; however, it has higher abrasion resistance. It is suitable for huge castings due to the minimum residual cooling stresses. Its parts require reasonable toughness; however, superior abrasion resistance like earth moving machines would choose leaner alloys.

Tool Steels

Tool steels are high carbon steels having high hardness, strength, and wear resistance. With 0.7% to 1.5% carbon content, tool steels are developed for forming and cutting operations. Alloying elements, such as chromium, tungsten, molybdenum, are added to improve the hardness and wear resistance of tool steels. The properties of tool steels can also be improved by heat treatment. Tool steels are prepared for various grades for diverse applications. American Iron and Steel Institute (AISI) and the Society of Automotive Engineers (SAE) have combinedly introduced a single-letter classification for tool steel. Out of them, grade O, D, T, M, and H tool steels are used for tribological applications [3]. Alloy composition, physical condition, and hardness of the tool steels are listed in Table **6**.

Table 6. Details of various tool steels [2 - 4].

AISI grade	Physical condition	Primary alloy composition (Maximum %)	Hardness (HB or HV)
W(1-3)	Water hardening	1.5 C, 0.4 Mn, 0.15 Cr, 0.25 Ni	531
O(1-7)	oil hardening/cold work alloy steel	1.5 C, 1.5 Si, 1.2 Mn, 0.3 Cr	185-577
A(1-10)	Air-hardening	5 Cr, 1.25 C, 1.0 Mn, 0.8 Mo	
D(1-7)	Diffused hardening/cold work alloy steel	14 Cr, 1.7 C, 1.0 Mn, 1.2 Mo	450-590
T(1-15)	High-speed tungsten tool steel	21 W, 5 Cr, 5 V, 1.6 C	832
M(1-62)	High speed molybdenum tool steel	4 Cr, 9 Mo, 2 W, 0.88 C	200-765
H	Hot work tool steel chromium-based (H1 to H19)	0.5 C, 5 Cr, 2 Mo, 1.5 Si	450-590
	Tungsten based (H20 to H39)	19 W, 0.05 C, 4 Cr-	
	Molybdenum based (H40 to H59)	0.5 C, 5 Cr, 5 Mo	

Cold Work Steels

These are high carbon steels with the alloying materials chromium, tungsten, and manganese. The O&D series come under this category. O series steels need to be oil quenched, whereas the D series steels need to be air quenched. These steels deliver great wear resistance at a low cost. The extreme working temperature for these steels is $188 - 223^0C$.

Chromium Hot-Work Steels

These are low carbon and more chromium steels with the alloying materials tungsten and vanadium. These are tough materials with hot hardness competence. Supersonic aircraft require combined wear resistance and toughness. H steel is useful for these kinds of extremely stressed parts.

High-Speed Tool Steels

These tool steels are intended for machining the workpiece at high removal rates. At high cutting speeds, tools must have a high hot hardness to withstand the effects of frictional heating. The T & M series steels are appropriate for this requirement. T series steels are composed of the highest tungsten and other compositions, namely molybdenum, chromium, vanadium, and cobalt. Carbon content in these steels ranges between 0.70 and 1.5. They are maximum wear-resistant tool steels, but they have low toughness. The M series is called high-speed molybdenum steels with the key element, molybdenum ranging from 4 to 9%. M series tool steels are cheaper and tougher than the T series.

Bearing Steels

Bearing steels are unusual steels containing extraordinary wear resistance and rolling fatigue strength. SS 440C is stainless steel that resists normal corrosion, which can be used for bearings. However, it is limited to low-temperature applications. 52100 bearing steel is one of the distinctive steel with the advantages of great wear resistance and rolling fatigue strength. High-carbon chromium-bearing steel and some stainless steel types can be used as bearings materials after heat treatment and surface treatment. Carburized steels are very much useful for roller bearings. AISI 4620 and AISI 4820 are useful steels for light and heavy bearing applications. Table 7 shows the details of various bearing steels and their properties.

Table 7. Details of various bearing steels and their properties.

Grade	Composition (Maximum %)	Density (kg/m³)	Thermal expansion (1/°K)×10⁻⁰⁶	Hardness (HB or VB)	Tensile strength (MPa)
52100	1C, 1.6 Cr, 1.25 Mn	6370	12.4	415	1379
50100	0.9 C, 0.6 Cr, 0.45 Mn	6370	12.4	225	689
51100	0.9 C, 0.9 Cr, 0.45 Mn	6370	12.4	200	689
4620	0.25 C, 1.75 Ni, 0.25 Mo, 0.55 Mn	7850	12.4	740	827
4820	0.25 C, 3.75 Ni, 0.3 Mo, 0.7 Mn	7860	15.4	690	1379
SS 440	0.75 C, 18 Cr, 1 Si, 0.5 Mn	7470	10.8	650	1379

NON-FERROUS METALS

Almost all metal alloys hold little, or insignificant quantity of iron in their composition. The metals that do not contain a significant quantity of iron in their chemical configuration are called non-ferrous metals. These metals, in general, contain iron of lower than one percent as measured by weight. Essential non-ferrous metals include copper, aluminum, lead, zinc, tin, gold, and silver. Their fundamental benefits over ferrous metals are their malleability and corrosion resistance.

COPPER ALLOYS

Copper base alloys have been used in industries for a long time due to their high corrosion resistance and good strength. Copper forms alloys more quickly than other metals, with an extensive range of alloying elements to attain useful metal. Copper alloys are majorly classified as brass, bronze, copper-nickel, gunmetal, and copper-beryllium. Table **8** shows the different copper alloys and their properties.

Brasses are copper and zinc alloys; they are strong enough and malleable. The brass's cold working properties can be improved by increasing zinc content by up to 35%. Brasses with 32% and 39% of zinc give outstanding hot working properties; however, they restrict cold workability. Like Muntz Metal, brasses with higher than 39% zinc consist of high strength and poorer ductility at room temperature compared to few zinc alloys. Brasses can easily casted, drawn, and formed to produce several components like springs, duct pipes, fire extinguishers, radiator cores, low-pressure gears, and low load bearings.

Table 8. Composition details and properties of various copper alloys [2 - 4, 17].

Alloy	Grade	Composition (%)	Tensile strength MPa	Hardness BH	Thermal expansion $(1/^0K) \times 10^{-06}$
Pure copper	C10200	99.95 Cu	221-455	50	16.5
Red Brass	C23000	85Cu, 15Zn	280	60	18.0
Muntz metal	C28000	60Cu, 40Zn	378	-	20.8
Phosphor bronze (or) Tin bronze	C51000, C52400	85-88Cu, 10-12Sn 0.2-0.3P, other	350-580	75	17.8
Aluminum bronze	C60800	95Cu, 5Al	420	150	17.5
	C63000	81.5Cu, 5Al, 5Ni, 2.5 Fe, 1Mn	700	170	16.2
Leaded bronze	C9320	81-85Cu, 6-8 Pb, 6-7Sn	310	65	20.3
Beryllium copper	C17200	97.9Cu, 1.9Be, 0.2Ni	490-1400	380	17.8
Copper Nickel	C71500	70Cu, 30Ni	385-588	120	16.2
Nickel silver	C75700	65Cu, 23Zn, 12Ni	427	175	16.2
high-leaded tin bronze copper	C93900	76-80Cu, 14-18Pb, 5-7Sn	172-221	63	16-19

Bronze is a copper alloy metal containing around 12–12.5% tin and additional metals like aluminum, manganese, nickel, or zinc. These accompaniments produce a variety of alloys, which are maybe harder than pure copper or possess other valuable properties, like stiffness, ductility, or machinability. Bronze was the hardest metal in widespread use during the bronze age.

Tin Bronze

Tin bronze is a hard and ductile metal with a diameter of 0.5% to 11% tin and 0.01% to 0.35% phosphorous. The existence of tin offers high mechanical properties. It has high load-carrying ability, wear resistance, and impact strength. This metal is frequently used in various tribological applications such as bearing, gears, bushing, and electrical contacts. Moreover, the alloys are famous due to their corrosion resistance in salt-water and brines.

Leaded Tin Bronzes

Fundamentally, lead and copper are immiscible in the liquid state. During solidification, the copper solidifies first (at 593 C). The lead solidifies at a much lower temperature (0 C) and is distributed in the copper matrix. There is a small solubility of lead in copper, which is about 0.003%. Lead is added to copper tin alloys to make leaded bronzes. It has up to 20% lead, which can be added to tin bronzes. The common alloys cover a 1 - 10% lead range. This alloy provides intrinsic solid lubrication due to the free lead in the metal. Lead at the mating shaft gives a soft coating that reduces friction and reduces overheating during lubrication oil loss. Leaded tin bronze bearings follow the rotor misalignment so that the interaction area is maximized. These benefits help distribute the load above the maximum possible surface.

Aluminum Bronzes

Aluminum bronzes are valuable due to their high strength, excellent corrosion resistance, and wear resistance properties combination. These alloys classically have 9-12.5% aluminum and up to 6% iron and nickel. In substitutional solid solution in the Cu crystalline lattice, aluminum delivers an enhancement of the mechanical strength of copper. It also improves the corrosion resistance, especially in acid surroundings, due to the establishment of a thin alumina film on the alloy surface.

Wrought aluminum bronzes consist of 4% to 12.5% of aluminum. If the alloy's aluminum percentage is less than 8%; the alloys are α-phase Al-bronzes and can be cold worked. The alloys with 8–12.5% aluminum with accompaniments of iron, manganese, nickel, and silicon are two phased (α–β) Al-bronzes. The adding of nickel is essential as it reduces de-aluminification.

Beryllium Copper

It is an uppermost strength copper alloy that exists commercially. It is also recognized as spring copper or beryllium bronze. The most common grades of beryllium copper contain 0.4 to 2.0% of beryllium. The age-hardened beryllium copper has the strength of near steel. It has high thermal conductivity and can be used at higher temperatures. It has outstanding corrosion resistance to naval and industrial situations. It is used for heavily loaded grease-lubricated bushings for bearing loads up to 3000 bar.

BABBITT METAL

Babbitt or white metal is one of the alloys used as a bearing surface in a plain bearing. Isaac Babbitt first created it in 1839, and it is the most regularly used bearing material. Bearing metals require a low coefficient of friction, even without lubrication. Babbitt bearing has much less friction than other metals like steel or cast iron. The coefficient of friction of the Babbitt metal bearings is even lower than ball bearings by adding lubrication. Table **9** shows the composition details of tin and lead-based Babbitt alloys and their properties.

Table 9. Composition details and properties of Babbitt alloys.

Babbitt	Grade	Composition (%)	Hardness (BV)	Tensile strength (MPa)	Thermal expansion $(1/^0K) \times 10^{-6}$
Tin base	ASTM B-23 (1,2,3,11)	80-92Sn, 4-8Sb, 4-7Cu, 0.35-0.5Pb	26	81	12-24
Lead base	ASTM B-23 (7,8,13,15)	1-10Sn, 14-16Sb, 0.4-0.5Cu, 70-85Pb	20	69	19.6-24

Tin-Based Babbitt

This alloy contains more than 80% of tin and lesser parts of antimony, lead, and copper. Lead-free tin-based Babbitt is used for the machinery that handles food. Even though Babbitt is a soft metal, the tin enhances its features' hardness to give more load-carrying capabilities. Manufacturers prefer tin-based Babbitt due to its wear resistance and low friction resistance. It can absorb dust elements from lubrication oil that are stuck on the metal surface. Tin-based Babbitt will not clutch or rub during metal surface contact. The mechanical properties of the tin Babbitt can be improved by adding cadmium. Tin-based alloys have a high thermal conductivity that allows them to move heat away and avoid hot spots. Due to the above advantages and high load capabilities, tin-based Babbitt bearings can be used in heavy machinery, motors, naval applications, *etc.*

Lead-Based Babbitt

It is one more common bearing material that contains 75% or more lead alloy and small amounts of tin and antimony. This alloy has outstanding corrosion resistance that makes it perfect for naval and high humidity and moisture applications. It also delivers low friction and can adhere to bronze and steel. Manufacturers generally use lead-based Babbitt metal in shock loads involvement applications. It is perfect for low and medium-speed equipment and bearings that experience cyclic loads. It is a cost-effective alloy as it can be used for common-purpose machinery and transmission systems.

ALUMINUM ALLOYS FOR BEARING

Aluminum alloy has a high strength-to-weight ratio, higher wear resistance, and corrosion resistance than classical bearing materials. They can be used at higher temperatures. These metals are cheaper and have fatigue strength and thermal conductivity. However, aluminum alloys' tribological applications are limited due to their surface properties and low wear resistance. Therefore, the addition of alloys and application of heat treatment are essential to achieve tribological and mechanical properties. Chromium, nickel, silicon, and tin are the few alloying materials.

Al-Si cast alloy is extensively used for engine components as an alternative for cast iron. The adding of silicon in aluminum alloys increases their wear resistance, casting, machining, and corrosion features. It is becoming the best alternative for cast iron in IC engines and other heat transfer applications. It has excellent thermal conductivity, which permits the combustion heat in engines to be taken out more quickly than cast iron.

Aluminum-tin alloy is an excellent bearing metal that is extensively used in IC engines and hydraulic pumps. It is a ductile material and can bear heavy loads without cracking. A higher amount of tin provides very good scratch resistance. However, as soon as the amount of tin is greater than 10%, it lowers the ductility and yield strength. An aluminum-tin-silicon-copper alloy is the most effective aluminum alloy metal for bearings due to its wear resistance and castability properties [18].

The most popular aluminum alloys for tribological applications are SAE 770, 780, and 781. SAE 770 (Al 850) is an aluminum-tin alloy used at low and medium shock loads with advantages of great fatigue strength and excellent corrosion resistance. The hardness of this metal is above the bronzes and below the Babbitt metal. The fatigue strength of this alloy is three times larger than tin or lead-based babbitts. Alloy 780 (Al 8280) is comparable to SAE 770, except for 1.5% extra silicon. Silicon addition increases strength and fatigue properties and also enhances wear resistance [19]. SAE 781 has 1 to 3% cadmium for better bearing features. It is readily used with steel baking in heavy load applications.

ZINC-ALUMINUM ALLOY

In zinc-aluminum alloy, zinc is the base metal, aluminum is a higher concentrated alloy, and the elements like magnesium and copper make up the other content. These are high-performance alloys that show decent strength, corrosion resistance, and hardness. ZA-8, ZA-12, and ZA-27 are common zinc-aluminum alloys, where the number denotes the aluminum percentage. ZA-8 and ZA-12 alloys are moderate to high strength materials, while ZA-27 is a high-strength alloy [20]. Table **10** shows their composition details and properties. These alloys have outstanding castability, corrosion resistance, and alike or even greater bearing and wear properties compared to the standard bronze bearing. These alloys are very much useful for various tribomechanical components in power plants, such as the radial and journal bearings, several bushings, guides, *etc.* Zn-Al alloys bearings have shown mainly good surface characteristics. Therefore, they are useful in locomotive motors and mining tribomechanical components at high loads and low speed circumstances.

Table 10. Composition details and properties of Zn-Al alloys.

Grade	Composition (%)	Hardness (BH)	Tensile strength (MPa)	Thermal expansion $(1/^0\text{K}) \times 10^{-6}$
ZA-8	8.8Al, 1.3Cu, 0.075Fe, 0.03Mg, balance Zn	103	263	23.3
ZA-12	11.5Al, 1.2Cu, 0.075Fe, 0.03Mg, balance Zn	89-100	276-317	24
ZA-27	25Al, 2Cu, 0.075Fe, 0.03Mg, balance Zn	119	425	26

NICKEL-BASED SUPERALLOYS

Nickel-based superalloys have both high strength and corrosion resistance in the course of work at higher temperatures. Aluminum and titanium are the important solutes in these alloys, with a total focus usually less than 10%. They have been extensively used in gas turbines and other power plants, where the working temperature is very high. Inconel 718 alloy that is largely used in the aircraft components is one of the nickel-based superalloys with the combination of nickel & chromium materials. It is one of the most useful alloys for metal additive manufacturing (MAM) [21]. Composition details are shown in Table **11**. As shown in Fig. (**8**), the microstructure has a matrix of γ grains and carbide boundaries. It has high-strength, corrosion-resistant, and can be used up to 700°C. A highly advanced and sophisticated application like aerospace, gas turbines, spacecraft, and nuclear reactors uses this material.

Table 11. Composition details and properties of nickel-based superalloys.

Composition (%)	Inconel 718	Inconel 625	Properties	Inconel 718	Inconel 625
Nickel	50-55	58 min	Density (kg/m³)	8230	8440
Chromium	17-21	20-23			
Iron	17	5	Ultimate Tensile strength (MPa)	1375	880
Niobium	4.75-5.5	3.15-4.15			
Molybdenum	2.8-3.3	8-10	Hardness (BHN)	<364	290
Other	1-4 (Co, Cu, Mn, Ti, C, P, S, Si)	0.5-2 (Co, Cu, Mn, Ti, C, P, S, Si)	Thermal Expansion $(1/°\text{K}) \times 10^{-6}$	11.5	12.5

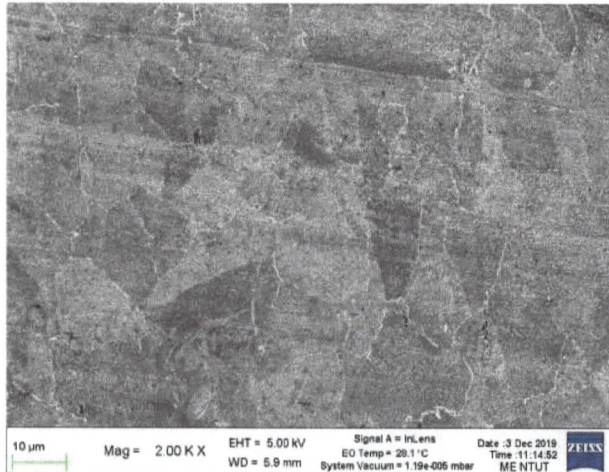

Fig. (8). Microstructure of Inconel 718.

Another nickel-chromium superalloy is Inconel 625, which has high strength, manufacturability, and exceptional corrosion resistance. As shown in Fig. (**9**), the microstructure contains a thick film of carbides along the grain borders. Working temperatures start from cryogenic to 982°C.

CM-247 LC is one of the nickel-based superalloys with a smaller amount of carbon content, which has advantages of carbide stability, mechanical and thermal properties, and ductility. It can be used in aerospace engines and turbines. Researchers are using laser-based surface treatments to enhance the tribological properties of this alloy [22].

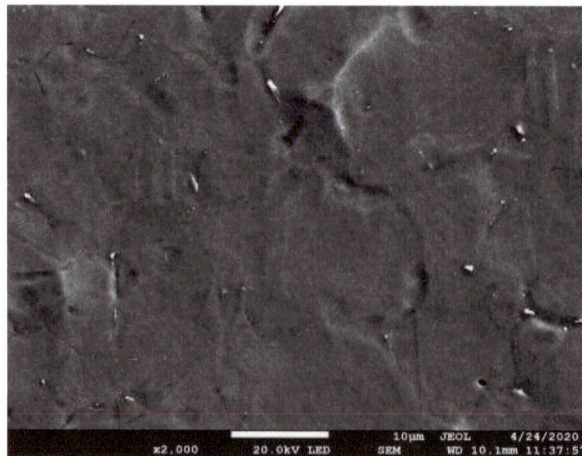

Fig. (9). Microstructure of Inconel 625.

Aerospace, car manufacturing, power generation, and chemical industries widely use titanium-based alloys due to their low weight and excellent corrosion properties. However, their applications are limited due to poor tribological properties. Surface treatment techniques are very much useful to overcome these limitations. For instance, AISI 420 alloy powder can be cladded on Ti–6Al–4V base material to improve the hardness and wear properties [23], NiCrMoNb and NiCrBSiC laser cladding can enhance the near-α titanium alloy properties [24].

CONCLUDING REMARKS

Machines can only work effectively if the design and material are suitable for the operating environment. Therefore, understanding the metal's behavior and selecting the appropriate metal are the essential steps in product development. Metals are broadly classified into ferrous metals and non-ferrous metals. Both classes have distinct advantages and disadvantages. Ferrous metals are harder but corrosive and heavy. In contrast, non-ferrous metals are soft, non-corrosive, and lightweight. The best metal can be achieved using various alloy elements, heat treatment techniques, and surface hardening processes. Machines in tribological applications experience a great extent of mechanical and thermal stresses. This chapter has explained the mechanical and physical properties of the existing metals and their applications.

CONSENT FOR PUBLICATION

Not applicable.

CONFLICT OF INTEREST

The author declares no conflict of interest, financial or otherwise.

ACKNOWLEDGEMENTS

This book chapter would not have been possible without the exceptional support of the book editor Dr. N. Jeyaprakash and the Bentham Science Publishers Pte. Ltd.

REFERENCES

[1]   Davis JR, Ed. ASM specialty handbook: cast irons. ASM international 1996.

[2]   Davis JR. Metals Handbook, desk ed. ASM International: Washington, DC. 1998; pp. 169-70.

[3]   Glaeser W. Materials for tribology. Elsevier 1992.

[4]   Cardarelli F. Materials handbook. London: Springer 2018.
      [http://dx.doi.org/10.1007/978-3-319-38925-7]

[5]   Taylor BJ, Eyre TS. A review of piston ring and cylinder liner materials. Tribol Int 1979; 12(2): 79-89.

[http://dx.doi.org/10.1016/0301-679X(79)90006-9]

[6]   Bahari A, Lewis R, Slatter T. Hardness characterisation of grey cast iron and its tribological performance in a contact lubricated with soybean oil. Proc Inst Mech Eng, C J Mech Eng Sci 2018; 232(1): 190-203.
[http://dx.doi.org/10.1177/0954406216675895]

[7]   Sun G, Zhou R, Li P, Feng A, Zhang Y. Laser surface alloying of CBW-Cr powders on nodular cast iron rolls. Surf Coat Tech 2011; 205(8-9): 2747-54.
[http://dx.doi.org/10.1016/j.surfcoat.2010.10.032]

[8]   Ngqase M, Pan X. An Overview on Types of White Cast Irons and High Chromium White Cast Irons. InJournal of Physics: Conference Series 2020; 1495(1): 012023.
[http://dx.doi.org/10.1088/1742-6596/1495/1/012023]

[9]   Jeyaprakash N, Yang CH, Duraiselvam M, Prabu G. Microstructure and tribological evolution during laser alloying WC-12% Co and Cr3C2− 25% NiCr powders on nodular iron surface. Results Phys 2019; 12: 1610-20.
[http://dx.doi.org/10.1016/j.rinp.2019.01.069]

[10]   Sánchez JC, Toro A, Estupiñán HA, Leighton GJ, Endrino JL. Fabrication of bio-inspired deterministic surfaces by photochemical machining for tribological applications. Tribol Int 2020; 150: 106341.
[http://dx.doi.org/10.1016/j.triboint.2020.106341]

[11]   Parreira JG, Gallo CA, Costa HL. New advances on maskless electrochemical texturing (MECT) for tribological purposes. Surf Coat Tech 2012; 212: 1-3.
[http://dx.doi.org/10.1016/j.surfcoat.2012.08.043]

[12]   Baharin AF, Ghazali MJ, Wahab JA. Laser surface texturing and its contribution to friction and wear reduction: a brief review. Ind Lubr Tribol 2016.

[13]   Seok J, Lee SO, Jang KI, Min BK, Lee SJ. Tribological properties of a magnetorheological (MR) fluid in a finishing process. Tribol Trans 2009; 52(4): 460-9.
[http://dx.doi.org/10.1080/10402000802687932]

[14]   Shiao Y, Kantipudi MB, Jiang JW. Novel Spring-Buffered Variable Valve Train for an Engine Using Magneto-Rheological Fluid Technology. Front Mater 2019; 6: 95.
[http://dx.doi.org/10.3389/fmats.2019.00095]

[15]   Jeyaprakash N, Yang CH, Tseng SP. Wear Tribo-performances of laser cladding Colmonoy-6 and Stellite-6 Micron layers on stainless steel 304 using Yb: YAG disk laser. Met Mater Int 2019 Nov 13: 1-4.

[16]   Ross RB. Metallic materials specification handbook. Springer Science & Business Media 2013.

[17]   Breedis JF, Caron RN. Wrought Copper and Wrought Copper Alloys. Kirk-Othmer Encyclopedia of Chemical Technology 2000.

[18]   Hunsicker HY, Kempf LW. Aluminum alloys for bearings. SAE Technical Paper 1947.
[http://dx.doi.org/10.4271/470201]

[19]   Shabel BS, Granger DA, Truckner WG. Friction and Wear of Aluminum-Silicon Alloy.

[20]   Seah KH, Sharma SC, Girish BM. Mechanical properties of as-cast and heat-treated ZA-27/graphite particulate composites. Compos, Part A Appl Sci Manuf 1997; 28(3): 251-6.
[http://dx.doi.org/10.1016/S1359-835X(96)00117-0]

[21]   Hosseini E, Popovich VA. A review of mechanical properties of additively manufactured Inconel 718. Addit Manuf 2019; 30: 100877.
[http://dx.doi.org/10.1016/j.addma.2019.100877]

[22]   Jeyaprakash N, Yang CH. Improvement of tribo-mechanical properties of directionally solidified CM-247 LC nickel-based super alloy through laser material processing. Int J Adv Manuf Technol 2020;

106(11): 4805-14.
[http://dx.doi.org/10.1007/s00170-020-04931-9]

[23]   Jeyaprakash N, Yang CH, Tseng SP. Characterization and tribological evaluation of NiCrMoNb and NiCrBSiC laser cladding on near-α titanium alloy. Int J Adv Manuf Technol 2020; 106(5): 2347-61.
[http://dx.doi.org/10.1007/s00170-019-04755-2]

[24]   Jeyaprakash N, Yang CH. Microstructure and Wear Behaviour of SS420 Micron Layers on Ti–6Al–4V Substrate Using Laser Cladding Process. Trans Indian Inst Met 2020 Mar 12 : 1-7.
[http://dx.doi.org/10.1007/s12666-020-01927-7]

CHAPTER 2

Tribological Issues with Metals

Milon Selvam Dennison[1,*] and **G. Vignesh**[2]

[1] *School of Engineering and Applied Sciences, Kampala International University, Western Campus, Uganda*

[2] *Karpagam Academy of Higher Education, Coimbatore-641 021, India*

Abstract: Metals and their alloys are widely utilized for engineering applications because of their enhanced strength and workability. When the metallic components are assembled to deliver the relative motion, the friction will be generated due to the interaction between the metallic surfaces, and this interaction will lead to the 'Wear' of the metallic components. Wear in mechanical components reduces the plant efficiency because of power losses, and in rare cases, it is catastrophic. Wear is a vital cause of malfunctioning a mechanical system, and it is a serious issue that needs to be addressed in tribological applications. In a mechanical system, the relative motion of a solid body/particles (or) fluid particles over other metallic surfaces results in fragmentation, plowing, cutting, shearing, scuffing, scoring, pitting, *etc.*, and these features can be categorized based on the wear mechanisms. The important wear mechanisms in industrial situations are abrasion, adhesion, delamination, erosion, fatigue, fretting and oxidation. This chapter comprehensively reviews the various tribological issues in the metals and also some notable case studies.

Keywords: Abrasion, Adhesion, Alloys, Cutting, Delamination, Erosion, Fatigue, Fragmentation, Fretting, Friction, Mechanical Components, Mechanical System, Metals, Oxidation, Plowing, Pitting, Shearing, Scuffing, Scoring, Wear.

INTRODUCTION

Tribology is a traditional scientific study, which deals with the mechanism of metallic or non-metallic surfaces interacting with each other. The terms wear, friction and lubrication relate to 'Tribology' [1 - 3]. In general, the need for learning tribology is to regulate the friction in the mechanical components. The tribological characteristics, such as wear and friction, can be deliberated with the aid of testing equipment named 'Tribometer'. The factors of the processes, such

* **Correspondence author Milon Selvam Dennison:** School of Engineering & Applied Sciences, Kampala International University, Western Campus, Uganda; Tel: +256707829516; E-mail: milon.selvam@kiu.ac.ug

Jeyaprakash Natarajan and Che-Hua Yang (Eds.)
All rights reserved-© 2021 Bentham Science Publishers

as ambient temperature, applied load, contact pressure, sliding speed, *etc.*, should be selected carefully for a significant tribological experiment of materials [5, 6].

It is well understood that engineering tribology is the study of wear and friction in materials concerning sliding contact. Suppose when two similar or dissimilar materials come in physical touch and sliding motion, the ratio value of the tangential friction (Ft) to the normal applied load (P) can be defined as a dimensionless scalar quantity known as the Friction Coefficient (or) Coefficient of Friction (µf) [3, 7]. The coefficient of friction depends upon the type of material and the nature of the applied load, and it ranges from zero to greater than one. The mathematical relation of the Coefficient of Friction is given in equation (1).

$$\mu_f = \frac{F_t}{P}$$ (1)

Fig. (1) illustrates the load applied to a body resting on a plane surface. If the applied load is more than the frictional force or resistance force, the body starts moving on the surface. The friction is defined as the resistance between the two surfaces in relative contact. The factors influencing friction are the surface roughness of the sliding material, material crystal structure, strain hardening, hardness, elastic and shear modulus, grain size, surface energy, normal load, sliding velocity, environmental and ambient temperature [7]. In detail, the prime factors affecting the wear resistance of the structural material causing intensive wear are as follows:

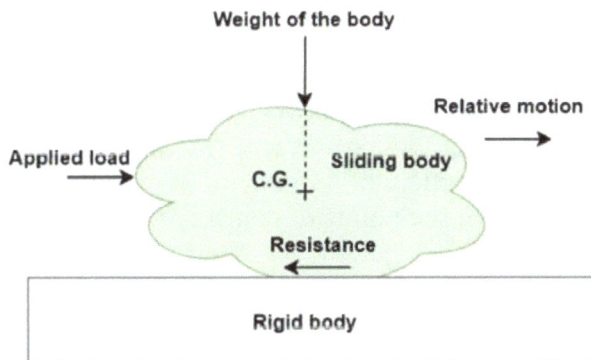

Fig. (1). A body sliding on a surface

• The properties of the sliding material and the quality of the rubbing surface.

- The nature of the movement (sliding, rolling, flow, impacts, cavitation, *etc.*).
- The temperature in the material interface zone and the magnitude of the load.
- The presence of lubricants and wear debris in the sliding interface.
- The presence of vacuum or corrosive environments that complicates wear processes.

The complexity of the wear processes has led to the creation of numerous methods for assessing the behaviour of the rubbing parts. It also led to the creation of an extensive class of test machines and standards. According to modern concepts, the external friction of materials has a dual nature, either molecular-mechanical or adhesive-deformation. It is believed that the interaction of materials due to the roughness or waviness of the surfaces occurs in separate zones of actual contact. The total area of these zones is called the actual or real area of the material being used. Therefore, the theoretical interpretation of the phenomena of wear is associated with significant facts, such as calculations and recommendations. Wear is always a system property, not a property of components involved. In scientific experimentation, the wear is deliberated in terms of 'wear rate,' which is the quantity of the material removed per unit time [2, 4, 6].

When analyzing the wear mechanism of the metal surfaces, the process of destruction or separation of the material and the accumulation of its permanent deformation is assessed depending on the types of friction, which are mentioned below [7, 8]:

- Static friction - friction of two bodies during micro-displacements before transitioning to relative motion.
- Friction of motion - friction of two bodies in relative motion.
- Friction without lubricant - friction of two bodies in the absence of any type of lubricant introduced on the friction surface.
- Friction with lubricant - friction of two bodies in the presence of any type of lubricant introduced on the friction surface.
- Sliding friction - the friction of motion of two rigid bodies, in which the velocities of the bodies at the points of contact are different either in magnitude or direction.
- Rolling friction - the friction of motion of two rigid bodies, at which their speed at the points of contact are the same in magnitude and direction.
- Friction force - resistance force during the relative movement of one body over the surface of another under the action of an external force directed tangentially to the common boundary between these bodies.

WEAR OF METALS

When two surfaces of the metals are in physical contact to transmit a relative motion, the continuous loss of metal in the form of debris due to friction is known as the wear of metals. Many studies over several decades have witnessed the issues of wear in metals. Various researchers have been carrying out the search for wear-resistant metallic components in recent decades. The past literature on the wear of metals and its alloys witness abrasion and adhesion are the major concerns, but even more issues, such as erosion, delamination, fretting, fatigue, and oxidative wear, are also viewed seriously. The four major wear mechanisms that act on a sliding body [8] are depicted in Fig. (**2**).

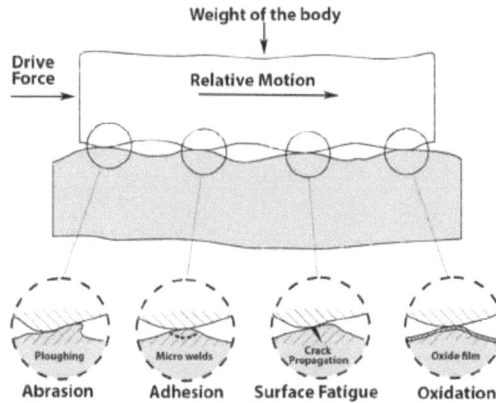

Fig. (2). Schema of major wear mechanisms.

Among the metals and their alloys, the wear caused by abrasion is most common in industrial situations, which cause surface damage to the power transmitting metallic components. In practical terms, the amount or degree of wear can be determined directly or indirectly. Some of the direct measurement methods are as follows:

    a. Determination of the absolute linear wear value in 'microns' or 'mm.'
    b. Determination of the absolute volumetric value of wear in 'μm^3' or 'mm^3.'
    c. Determination of the absolute value of wear by weight in 'g' or 'mg.'

As indirect characteristics of the wear process, the total duration of the occurrence of the phenomena or the wear time, the amount of material lost during friction, and other indicators can be used. Some of the factors influencing the wear of the metallic components subjected to cyclic loading are as follows:

- Design factors, such as shape, dimensions, clearances, and fit of the mating parts.
- Nature of the applied load (frequency of the load, shock, vibration, *etc.*).
- The interaction of the components - the type of friction, sliding speed, surface temperature, specific pressure, structure and hardness of the material, surface integrity, surrounding atmosphere, such as air, protective gas and vacuum.
- Intermediate medium - lubricant, the quality of the lubricant, surface films, abrasives (hardness of the abrasive, shape and size of the abrasive particles).
- The quality of the material used for the manufacture of parts, mechanical and heat treatment of parts.

WEAR MECHANISMS OF METALS

To analyze the wear resistance of materials, which is usually described by the wear rate, and to understand the mechanism of wear of friction pairs, more data are needed to observe the changes in them caused by plastic deformation, temperature rise, and chemical action or other factors. Depending on the state of the material, the degree of wear can be expressed as the loss of mass of the material, its deformation, material transfer and changes in properties. The summary of the various wear mechanism of metals, definition, and features are given in Table **1**.

Table 1. Summary of wear mechanism of metals, definition, and features

Mechanism	Definition	Features
Abrasion	When a harder tough metallic surface slides over a relatively soft metallic surface, the experienced wear in the form of fragmentation, plowing, and cutting is known as 'Abrasion.'	Fragmentation, cutting, and plowing
Adhesion	Adhesive wear is defined as the corrosion of metals by mechanical action. The other terms are scuffing wear or sliding wear.	Shearing, relocation of material, adhesive bonding, plowing
Delamination	The wear due to the weakening of the coated metal surface layer due to the loss of bonding is known as delamination.	Coating strength, durability, material aesthetics, crack propagation, and nucleation
Erosion	The wear on a metallic surface due to the turbulence of the impinging solid or fluid particles is known as erosion.	Pitting, turbulence, surface Deformation
Fretting	Fretting is the wear due to the oscillatory motion of the metallic interfaces. It is also called chafing corrosion.	Vibration, rubbing action, relative motion, deformation
Fatigue	It is the fracture aggravated in the metal due to the effect of the continuous loading or vibration in the metallic components.	Fatigue, cyclic stress, crack propagation

(Table 1) cont.....

Mechanism	Definition	Features
Oxidative	This is a type of wear resulting from the sliding action between two metallic components that generate oxide films on the metal surfaces. These oxide films prevent the formation of metallic bonding among the sliding metal surfaces, which results in the form of fine wear debris.	Oxidation, environment, interaction of metals

Abrasive Wear

Wear in metallurgical science is defined as the dislocation of the material from its original shape. Let us consider a harder material kept in proximity with a relatively softer material to transmit relative motion. The asperities of the harder material impose a plastic flow over the softer material; the wear caused by such a mechanism is termed as 'Abrasive wear' [9, 10]. Figs. (**3** and **4**) depict the schema of the abrasive wear mechanisms.

The abrasive wear can be categorized into two-body abrasion and three-body abrasion based on the contact environment and the type of contact. Two body abrasion refers to the gliding surfaces of harder and softer materials, where the harder material plows the softer material. For example, a file is used to shape a workpiece material in a mechanical workshop. Three-body abrasion refers to the process of removing materials with the aid of particles between the surfaces of the materials, for example, tumbling a finishing process in the production of a component.

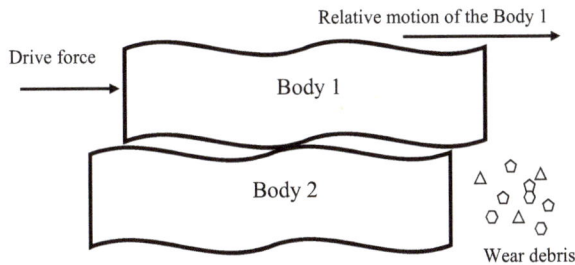

Fig. (3). Two-body abrasion.

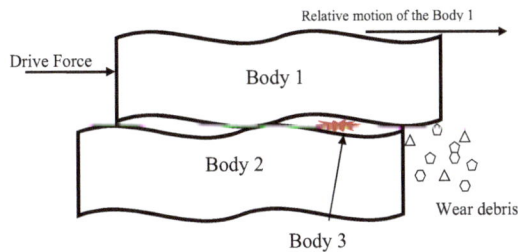

Fig. (4). Three-body abrasion.

There are so many factors that influence the incidence of abrasion, and several mechanisms define the material removal. The major mechanisms that relate to abrasion are as follows:

Fragmentation

Fragmentation happens when a material is partitioned by the cutting process that results in abrasiveness in its surface. This abrasiveness causes localized fracture of the material, and these fractures propagate freely around the wear groove, resulting in more removal of material by spalling.

Plowing

Plowing happens when the material displacement is in the sideways direction away from the particle of wear. The result is the formation of the groove, and it does not involve in the direct material removal process. This displaced material forms ridges adjacent to the grooves, which may be further removed by consequent abrasive material passage.

Adhesive Wear

Adhesive wear occurs when the surface of the two materials is in sliding contact. The atomic forces between the surfaces of the material interface become stronger. The shear begins in the weaker material, and the wear particles adhere to the stronger material surface [11]. This type of wear depends upon the chemical, physical and dynamic factors, such as the presence of chemicals (or) corrosive agents, material properties, sliding velocity, and even the applied load.

Fig. (5) illustrates the wear mechanism schema of adhesive wear on metals. Consider two metallic surfaces that come into gliding contact. The initial proximity of the metal surfaces could be observed in a few rough spots where the friction and the wear originate. When a driving force is applied to one of the metallic components, the rough spots in the material surfaces deform plastically and bind together due to high pressure and temperature. As gliding continues, the binding between the surfaces breaks and creates a cavity in one metal surface and depression on the other surface. The detached abrasive particles rub against the surface resulting in adhesive wear (or) adhesion. The major kinds of adhesive wear are sliding wear, galling wear and scuffing/scoring wear. Adhesive wear can cause serious problems, such as scoring, cold welding, pits, built-up edges, seizing, tool breakage and scuffing in the metallic components.

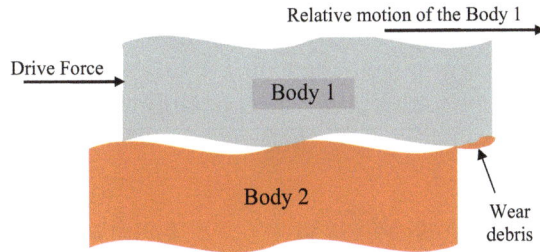

Fig. (5). Schema of adhesive wear.

Adhesive wear is the most important type of wear for most unlubricated wear systems or systems in partially lubricated conditions. One speaks of cold welding or seizure when small connections occur between the friction partners that are torn out of the surfaces due to further relative movement to one another. These torn parts leave small holes in the surface. The most important measures to prevent adhesive wear without lubrication are

- Selection of materials with softer oxides, *i.e.*, 'oxide lubrication.'
- By increasing the hardness of the metal surface to prevent the 'micro-plastic' deformation of the surface.
- Selection of highly incompatible metal pairings (such as silver on cobalt).
- Selection of metals with low surface energy.

Delamination Wear

The theory of delamination wear was first postulated by Suh in 1973, and this kind of wear is occasionally referred to as surface fatigue wear [12]. Delamination wear usually occurs in laminated materials and reinforced structures. Delamination tends to devastate the material coating strength, the aesthetic appearance of the material and its durability.

Nowadays, the use of layered materials is increasing in the aviation sector. As an air safety concern, the delamination wear occurs in the tail section of the airplanes. The reason for delamination wear is weakening in bonding. Thus, such wear is a menacing type of failure as it develops in the inner section of the material without being visible on the material surface, which is similar to metal fatigue. Delamination wear takes place in the following three phases:

a. Formation of voids near the surface
b. Further growth in the voids leads to crack propagation
c. When the crack propagation reaches a critical distance, the wear particles detach and break.

The formation of voids is usually observed in the materials with imperfections. Metal impurities are an important reason for the nucleation of these voids. Thus, materials with fewer impurities suffer less delamination wear.

Erosion Wear

Erosion wear is the damage experienced to a metallic surface when solid or fluid particles intrude on the surface. Based on the particle flow, it can be classified as solid erosion and fluid erosion [7, 13].

The surface contact stresses arise when the solid particles bump into the surface with greater velocity resulting in solid erosion, and it is one form of abrasion. The size of the solid abrasive particle, the velocity of the particle, and the impact angle are the factors to be considered in solid erosion. The continuous impact of the abrasive particles leads to wear debris in the metal. The wear rate in the case of solid erosion depends on the impact angle for brittle and ductile materials (Fig. **6**). For brittle materials, the material removal takes place by the formation of cracks at the intersection that exude from the point of impact of the eroded particle. Sharper abrasive particles lead to more deformation and wear as compared to round particles. The ductile material undergoes erosion by the plastic deformation process and the maximum erosion occurs at the angle of 20^0 (approx.) [7, 13].

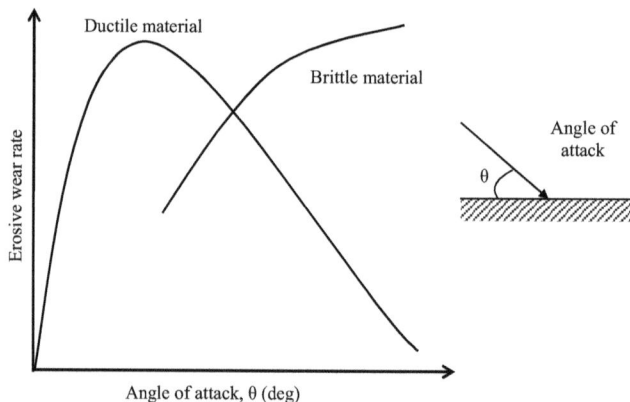

Fig. (6). Rate of erosive wear as a function of attack angle of the intruding particles for ductile and brittle materials.

Fluid erosion is of two kinds as fluid impact erosion and cavitation erosion. When minute liquid drops hit on the solid surface at high velocity and pressure, the material undergoes plastic deformation, and the repeated impacts lead to pitting and erosion. The rate of erosion increases in turbulent fluid flow and causes serious leakage in pipes. The metallic pipes (or) tubes usually have a protective flimsy layer, which is the initial part to be eroded by the fluid. Once the layer gets eroded, the bare metal surface is exposed to corrosion. Cavitation erosion is a special type of wear caused by water bubbles produced by pump impellers. This cavitation results in the formation of pits on the metal surface. Erosion can also happen due to poor workmanship. When the burrs in the inner surface of the tubes are not removed during the installation process, these inner burrs create turbulence and hamper the smooth running of the fluid, which finally leads to pitting in the metallic tubes [7, 13].

Fretting Wear

Fretting wear is the critical wear in mechanical design, and also, there is no standard method to identify fretting. The problems created by fretting wear are expensive to fix. Fretting wear is very common in machine components that are exposed to vibration [14]. In machine components, this type of wear occurs whenever the joints are subjected to oscillatory motion in the tangential direction with minimum amplitude. Fretting wear can also be understood in the way of adhesion (or) abrasion. The oscillatory motion results in rupture, creating wear debris, whereas the normal load results in adhesion between asperities. Also, the sliding of the machine components creates vibration and results in fatigue failure. This form of failure is called fretting failure. Fretting in the corrosive environment creates wear debris, which is harder than the parent metallic components and this can lead to abrasion. The truth to be known is, macroscopic gliding between the fretting contacts is minimal or negligible so that the fretting wear debris (130 μm) is supposed to be trapped between the contacts. Therefore, the wear rate per unit gliding due to corrosive fretting is relatively more than the adhesive and abrasive wear. The contacts which are subjected to fretting wear have a characteristic appearance of reddish-brown color and the debris is often referred to as 'cocoa.' In the case of ferrous metal fretting, the metal surface has a characteristic appearance of red/brown patches and the corresponding spots are highly polished to retrieve the quality [2, 4, 14].

Fretting wear on the metal does not have a single solution, as the main cause is oscillatory mechanical action (vibration). This type of wear might be overcome by eliminating the vibration in the interacting parts employing lubrication. Meanwhile, the presence of oxygen at the interacting part leads to the formation of iron oxide fretting. This can be prevented by scaling the spot, and through several coatings, such as phosphate, tufftride (or) sulphinuz coating prevent the adhesion between the parts, and this reduces fretting [2, 4, 14].

Fatigue Wear

Fig. (7) shows the schema of fatigue wear on metals. Fatigue wear occurs due to the application of alternating loads. Fatigue failure of the surface manifests itself in the form of growing microcracks [15]. Fatigue wear can begin deep in the metal with subsequent appearance to the surface. The few components which are most susceptible to fatigue wear are crankshafts, bolts in the connecting rod, rotors, springs, gears and valve plates. Fatigue wear appears in places of stress concentration, which can be used to trim burrs, inhomogeneities of forgings, sharp edges and methods of plastic deformation (for example, rolling with rollers), *etc.*

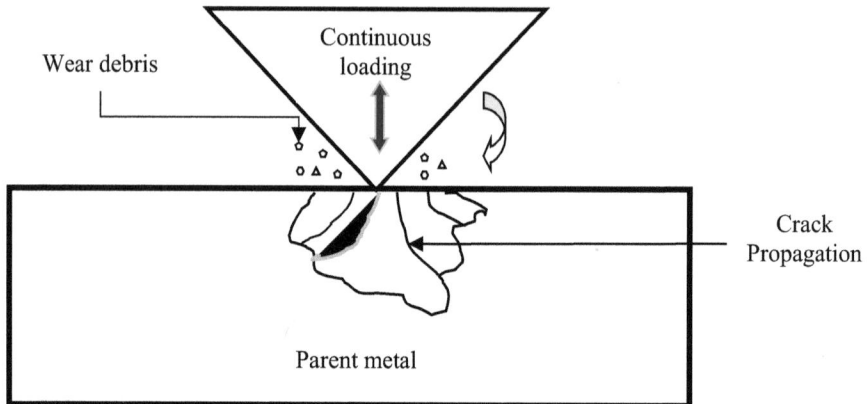

Fig. (7). Schema of fatigue wear.

Wear from the effects of power loads, such as torsion, bending, and impact manifests itself in the plastic deformation of parts. The components, such as shafts, rotors, keyed and spline connections, coupling pins and other parts under the action of loads, can change the shape of the working surfaces. Under loads higher than the calculated ones due to poor assembly or as a result of shock loads, connection parts such as keys can be not only deformed but also sheared, the bolts are lengthened and the thread profile is distorted. Thermal wear also manifests itself in plastic deformation and is associated with metal creep. For carbon steels, creep occurs at temperatures above 375 $^{\circ}$C, for alloyed steels, at more than 420 $^{\circ}$C [2, 3, 7].

Oxidative Wear

Corrosion is the process of converting metal into an oxidized state. As a result of oxidation, metals lose electrons and form oxides, hydroxides, or salts. According to the mechanism of interaction of metals with the environment, chemical and electrochemical corrosion are distinguished. Tribo-oxidation occurs when protective layers are removed by wear in a corrosive environment.

Oxidative wear is a sign of the structural variation of metals to friction [16]. This type of wear proceeds under normal friction conditions, and it consists of the formation of secondary films of solid solutions and chemical compounds of metal with oxygen and their removal from the surface of rubbing parts. The main feature of oxidative wear is the rate of destruction of films that do not exceed the rate of its formation. Plastic deformation activates the surface, stimulating the oxidation process; deformation and oxidation processes occur in the surface layers about 10 nm to 100 nm thick. Oxidative wear may occur in the metallic component of normal or elevated temperatures, without lubrication or when it is insufficient, and also in the absence of an aggressive environment. The wear rate during oxidative wear is considered to be low due to the protective effect of oxide films that prevent surfaces from seizing. Under the influence of repeated plastic deformation, the films maybe damaged, and if the films are solid, then their particles can be abrasive, scratching the surface. Damaged oxide films are very quickly restored, while the process of film reduction is enhanced under the influence of temperature in the friction zone.

Oxidative wear manifests itself at the following sliding speeds: under dry friction, for annealed steels, up to 4 m/s and for hardened steels, up to 7 m/s, with boundary friction up to 25 m/s. In this case, the specific load should not exceed the values at which the oil films or oxide layers are destroyed. The temperature in the friction zone is limited to 200^{0}C; at a higher temperature, the metal softens and the lubricant desorbs from the surface [17 - 19].

CASE STUDY: DRY SLIDING OF TITANIUM ALLOY

In recent years, titanium alloys (Ti-alloys) have been extensively used in automobile and aerospace industries due to their low density-to-strength ratio, which exhibits significant energy savings. The gas turbine elements, such as drive shaft, flange couplings, coupling fasteners, *etc.*, play a vital role in aerospace industries and hence the anti-wear property of these elements is highly significant for acquiring enhanced life span [20]. A list of Ti-alloys employed in various engineering applications is summarized in Table **2**. Even though, Ti-alloys have high strength and fracture toughness, the tribological properties like friction and wear resistance are poor [21]. The poor anti-wear-ability of Ti-alloys was attributed to lower work-hardening capability, minimal tribo-oxide protection and poor resistance against plastic deformation [22]. The Ti-alloys applied for tribological applications exhibits huge wear loss in tribo-contacting asperities due to permanent deformation, propagation of crack, and fracture. For improving the wear resistance properties of Ti-alloys, there are many surface alteration techniques, such as dimple-texturing, laser surface alloying, physical vapor deposition, plasma nitridation, thermal and micro-arc oxidation, can be applied on its tribo-contact surfaces. Poor wear resistance will be the cause of the development of rupture and shear adhesive junctions because of peak local tribo-contact pressure during rubbing. Generally, the Ti-alloys experiences two kinds of wear, oxidative wear and delamination wear at different temperatures, applied loads and sliding velocities. The responsible factors for the wear mechanism are identified to be adhesion, abrasion, material transfer, layer delamination and mechanical alloying [23].

Table 2. Composition and hardness of various Ti-alloys.

Ti-alloy grade	Composition (wt%)	Hardness, HRC
TC4	5.65Al, 4.02V, 0.15O, 0.062C, balance Ti	30
TC5	6Al, 4V, balance Ti	32
TC6	6Al, 2Sn, 4Zr, 2Mo, balance Ti	33
TC7	20Zr, 6.5Al, 4V, balance Ti	33
TC9	10V, 2Fe, 3Al, balance Ti	34
TC11	6.50Al, 3.54Mo, 1.57Zr, 0.32Si, balance Ti	34

The rubbing wear of Ti-alloys can be defined by the progression of the phenomena's like permanent deformation in the surface and subsurface, development of wear debris and material transfer, mechanical mixing and chemical reaction with the environment [24]. Wear resistance and wear behavior of Ti-alloys are evaluated by rubbing conditions, counter-face materials and reaction between tribo-contact surfaces and the environment [25, 26]. Based on Archard relation, the measurement of dry sliding wear is given in equation (2).

$$W_V = C_W \frac{P \times d}{H_D} \tag{2}$$

Where W_V is the wear volume, P is the load applied, d is the linear sliding distance, H_D is the softer material hardness of the two mating elements, and C_W is the wear coefficient depending on the material. When the wear coefficient (C_W) is less than 10-4, it could be treated as mild wear and is given in equation (3).

$$C_W = \frac{M_C \times k_C}{S_V \times O_T} \tag{3}$$

Where M_C is a material constant, S_V is the sliding speed, O_T is the tribo-oxide thickness, and k_C is the constant for oxidation rate, which is given in equation (4).

$$k_C = A_C \left(-\frac{A_E}{R \times F_T} \right) \qquad (4)$$

Where A_C is the Arrhenius constant, A_E is the activation energy for oxidation, R is the gas constant, and F_T is the surface flash temperature [26]. In the dry sliding of Ti-alloys, the first body has tribo-contact. In tribo-contact interfaces, the first body represents the sliding material (Ti-alloys), the second body represents the counter-face materials and the third body represents the region with mixed compositions from the first and second bodies. The coefficient of friction of the tribo-contact surface gradually decreases, resulting in material substrate peel-off. The peeled-off material will further grain to form wear debris and involve the dry sliding as a third body [27]. The tribo-oxide layer can act as a third body, which is obtained from the wear debris of the tribo-contact pairs, and it would develop a worn-out surface in the first and second bodies. Additionally, some amount of wear debris develops at the tribo-contact interfaces. Hence, the third body region contains the tribo-oxide layer on the worn-out surfaces of tribo-contact pair materials and some amount of wear debris retains in the tribo-contact interfaces [28]. Table **3** summarizes the recent works on the dry sliding wear of Ti-alloys.

Table 3. Summary of recent works on dry sliding wear of Ti-alloys

Authors	Sliding Material	Counter-face Material	Sliding Conditions	Temperature	Tribo-oxides	Tribology test type
Sahoo *et al.* [20]	TC5	EN 31 Steel	50–150 N 0.3 to 0.9 m/s	Room Temperature	TiO	Pin-on-disc
Wang *et al.* [21]	TC11	AISI 52100 steel	50 to 250 N	298 K to 873 K	TiO_2 & Fe_2O_3	Pin-on-disc
Li *et al.* [22]	TC11	AISI 52100 steel	10 to 50 N 0.5 to 4 m/s	-	TiO_2	Pin-on-disc
Farokhzadeh *et al.* [23]	TC5	AISI 52100 steel balls	0.8 N to 10 N	Room Temperature	MMC	Pin-on-disc
Wang *et al.* [24]	TC11	AISI 52100 steel AISI M2 steel	50–250 N	25–600°C	TiO, TiO_2, Fe_2O_3	Pin-on-disc
Chen *et al.* [25]	TC11	AISID2 steel AISI 52100 steel	50–200 N	25–600°C	Fe_2O_3, Fe_2TiO_5, & TiO_2	Pin-on-disc
Straffelini *et al.* [26]	TC5	Steel	0.2 to 1 m/s	Room Temperature	Ti & TiO	Pin-on-disc

(Table 3) cont.....

Authors	Sliding Material	Counter-face Material	Sliding Conditions	Temperature	Tribo-oxides	Tribology test type
Zhong et al. [27]	TC7	AISI 440C stainless steel ball	10 N, 30 N	800 °C, 900 °C & 1000 °C	TiO	Pin-on-disc
Zhou et al. [28]	TC5	AISI M2 steel AISI 52100 steel	50–250 N	25–600°C	TiO_2, TiO_2, & $Ti_{2.5}O_3$	Pin-on-disc
Wang et al. [29]	TC11 TC5	AISI 52100 steel	50–250 N	5–600°C	Fe_2O_3	Pin-on-disc
Cui et al. [30]	TC5	GCr15 steel	50–250 N	20–400°C.	Ti_8O_{15} & TiO_2	Pin-on-disc
Chen et al. [31]	TC11	AISI 52100 steel	50–250 N	25–600°C.	Fe_2O_3	Pin-on-disc
Mao et al. [32]	TC5	GCr15 steel	50–250 N	25–500°C	Ti_8O_{15} & TiO_2	Pin-on-disc
Li et al. [33]	TC5	AISI 52100 steel	0.5-4 m/s	Room Temperature	TiO, TiO_2, Fe_2O_3	Pin-on-disc
Heilig et al. [34]	TC6	AISI E52100 steel	5 and 10 N	800°C	TiO_2	Pin-on-disc
Zhang et al. [35]	TC5	AISI 52100 steel	10–50 N	Room Temperature	Fe_2O_3 and Fe_2O_3-rich	Pin-on-disc
Zhang et al. [36]	TC11	AISI 52100 steel	200 N	25 and 600°C	TiO_2, Fe_2O_3 & Fe_2TiO_5	Pin-on-disc
Hadke et al. [37]	TC5	Steel disc	10–40 N 200 - 800 rpm	Room Temperature	TiO_2	Pin-on-disc
Li et al. [38]	TC11	AISI 52100 steel	10 to 50 N 0.75–4 m/s	Room Temperature	TiO, TiO_2, Fe_2O_3 & Ti_7O_{13}	Pin-on-disc
Kumar et al. [39]	TC5	Polycrystalline alumina disc	13.7, 68.7 and 109.9 N 0.01, 0.1 and 1 m/s	Room Temperature	Al_2O_3	Pin-on-disc
Jeyaprakash et al. [40]	TC5	Hardened WC balls	35 N, 0.5 m/s	Room Temperature	MMC	Flat-on-disc
Heilig et al. [41]	TC6	AISI E52100 ball	5 and 20 N	Room Temperature	MMC	Flat-on-flat
Zhong et al. [42]	TC7	AISI 440C steel	0.39 to 1.17 m/s	Room Temperature	MMC	Ball-on-disc
Zhang et al. [43]	TC4	AISI 52100 steel	1 to 4 m/s	Room Temperature	Fe_2O_3	Ball-on-disc

(Table 3) cont.....

Authors	Sliding Material	Counter-face Material	Sliding Conditions	Temperature	Tribo-oxides	Tribology test type
Yang *et al.* [44]	TC7	AISI 440C balls	10, 20 and 30 N 300 to 600 rpm	Room Temperature	MMC	Ball-on-disc
Liang *et al.* [45]	TC7	quenched 45# steel	25, 50 and 100 N	Room Temperature	MMC	Ball-on-disc
Zhang *et al.* [46]	TC7	GCr15 ball	25, 50 and 100 N	Room Temperature	MMC	Flat-on-flat

Mechanism of Tribo-Oxide Layers

During the preliminary stage of rubbing, some fragments separate from the worn-out surfaces of Ti-alloys due to delamination and/or adhesion. The separated fragments further oxidize and entrap to produce a tribo-layer, or they escape from the tribo-contacts to form wear debris. Severe abrasive, adhesive wear and formation of wear debris can be observed in the preliminary stage of dry sliding at the tribo-contact surfaces of the Ti-alloys. Some fragments of wear debris separated from the sliding interfaces may retain. During continuous sliding, the retained metal wear debris might be agglomerated and deposited in the worn-out surfaces. Those wear debris further sintered to form a solid layer on the Ti-alloy matrix, which is named as tribo-oxide layer [31]. At the worn-out surfaces of the Ti-alloy, the formation of the tribo-oxide layer can act as a solid lubricant to avoid metal to metal contact and significantly influence the load-bearing capacity [22].

During rubbing, the heat generated at the tribo-contacts and external heat accelerate the chemical reaction between oxygen in the ambient air and delaminated layers of worn-out surfaces resulting in the development of the tribo-oxide layer. Three kinds of tribo-layers form on the worn-out surfaces of the Ti-alloys, the transfer layer (TL) contains a rich quantity of oxygen, the mechanically mixed layer (MML) contains a medium quantity of oxygen and the composite layer (CL) obtains a very low quantity of oxygen. The TL and CL form at elevated friction temperatures, and MML forms at room temperature, which depends on the sliding velocity and load. Compact and thick tribo-oxide layers completely avoid the metal-to-metal interaction and create protection against adhesive and abrasive wear. Due to centrifugal force, some quantities of wear debris could be completely removed from the tribo-contact interfaces [43].

The freely moving wear debris is identified as a third-body abrasive to maximize the wear rate, while non-moving wear debris agglomerates develop a tribo-oxide layer. It must be understood that tribo-layers and tribo-oxides should be distinguished because these two are usually confused concepts. The tribo-oxides are chemical reaction products between wear debris and oxygen present in the ambient air during the rubbing of Ti-alloys. The tribo-layers are mutually developed by retained wear debris and tribo-oxides [33]. The development of tribo-oxide layers is an unavoidable process during the rubbing of Ti-alloys that influences the wear mechanism and behavior. The tribo-oxide layers could be developed by the following techniques: at first, wear debris produced in the tribo-contact surfaces, in which some of the wear debris can be removed from the contact interface and other retained in the worn-out surface and oxides get off. Finally, the stored wear debris is compacted and sintered to develop the tribo-oxide layer [33]. Sintering is a diffusion process, whereas molecules bond together by atomic transfer at temperatures below the melting point [35].

The development of tribo-layers in the Ti-alloy contact surfaces can be controlled by rubbing conditions, in turn, the tribo-oxide layer decides the amount of wear rate. The developed tribo-layers do not contain pure tribo-oxides and might be composed of pure wear debris or mixed form of wear debris and oxides. Various forms of tribo-layers will exhibit different influences on the wear behavior and mechanism. It shows that the number of tribo-oxides present in the tribo-layer might be the key factor for determining the wear rate of Ti-alloys. The large surface areas of the tribo-oxide layer and small delaminated fragments exhibit a minimal wear rate. The smooth, fine and dark appearance of the worn-out surfaces shows the development of the tribo-oxide layer. The formation and activation of the tribo-oxide layer mechanism during dry sliding contact of Ti-alloys include (1) chemical affinity towards oxygen present in the ambient air, (2) availability of free thermal energy, (3) mechanical stability of the tribo-oxide. Ti-alloys have a great influence on the oxygen present in the ambient air, resulting in a huge charge difference [39].

With the presence of the tribo-oxide layer at the worn-out surfaces, the surface morphology could be altered to typical oxidative layer morphology, *i.e.*, fine, tribo-oxide zone and delaminated layer zone. This represents the tribo-oxide layer forms ceramic-to-ceramic contact instead of metal-to-metal contact at the tribo-contact surfaces of the Ti-alloys [36].

Influence of Temperature on Dry Sliding Wear of Ti-Alloys

During the sliding of Ti-alloys with the counterpart, materials generate flash temperature at the tribo-contact interfaces, and it leads to the oxidation of worn-out surfaces. The heat generated at the tribo-contacts and the external heat influence the thermal softening of the subsurface matrix. The thermal softening of the subsurface matrix results from supplementary tempering and/or the recrystallization under the combined action of permanent deformation and temperature. Furthermore, the tribo-oxide layer formed on the worn-out surfaces and the number of tribo-oxides directly depends on the amount of heat generation. The high temperature at the tribo-contact surfaces can alter the tribological and mechanical properties of rubbing surfaces, thus influencing the wear behavior [24]. The continuous and strong tribo-oxide layer produced at the elevated temperature significantly expose different properties than that of Ti-alloys [36].

At lower temperatures, the formation of the tribo-oxide layer would be in a scattered manner and discontinuous form, therefore, it cannot influence the wear resistance significantly. But if the tribo-contacts temperature reaches between the range of $500^{\circ}C$ to $600^{\circ}C$, then the number of tribo-oxides will be more in the tribo-oxide layer, which influences the high thick protective layer and high hardness in the worn-out region, resulting in the betterment in the wear resistivity of the Ti-alloys [21]. Even though an elevated temperature at the contact pairs influences the production of more oxides at the worn-out surfaces, an elevated temperature relaxes the sub-surface layers under the tribo-layers and results in the delamination of tribo-oxide layers. Ti-alloys are the best suitable for high-temperature rubbing parts because they avoid deformation and can withstand their original dimensions during rubbing at high temperatures [29]. When there is a rise in contact temperature at the tribo-contact, it will increase the number of tribo-oxides and wear debris on the tribo-contact surfaces. This will induce the formation of a thick and compact tribo-oxide layer [38].

At lower sliding temperature, typical adhesive and abrasive wear characteristics like plastic deformation, grooves and ragged traces appear on the tribo-contact points and much fine metal wear debris produces in the wear progression. There is an extensive wear rate in the Ti-alloys due to the cause of lower work-hardening ability and permanent shearing resistance [25]. In the case of the abrasion wear stage, one portion of free thermal energy would be utilized from the whole free thermal energy to plow, cut and distort the material. Here, the distorted wear debris will induce the work hardening process on the tribo-sliding surfaces and minimize the wear rate significantly [40]. No oxide tribo-layer at a lower temperature was commonly attributed to non-favorable Pilling-Bedworth ratio and discrepancy in thermal expansion among tribo-oxides and subsurface matrix [36]. During the sliding of Ti-alloys at room temperature, Ti-alloys expose severe wear at the tribo-contact points and result in a permanent deformation matrix. Hence, Ti-alloys are categorized to expose poor resistance against plastic shearing and low strength hardening ability, resulting in poor wear resistance.

Influence of Microstructure on Dry Sliding Wear of Ti-Alloys

The micro-structural modifications significantly influence the abrasive wear resistance of Ti-alloys at the tribo-contact surfaces [37]. The mechanical and tribological properties of Ti-alloys mainly depend on the micro-structural topographies [20]. The nanocrystalline structure of the tribo-oxide layer contains soft and hard micro-constituents being the rate of wear reduction. The oxidation and adhesion wear significantly increase with the material transfer on the surface topography [34]. This phenomenon shows that as far as tribo-oxide appears on the worn-out surfaces of the Ti-alloys, the wear rate would significantly reduce in any sliding conditions [36]. It clearly shows that tribo-oxide layers exhibit anti-wear protection against dry sliding of Ti-alloys. The wear protective function of tribo-oxide layers does not depend on its strength, and it purely depends on binding strength with the Ti-alloys subsurface matrix.

In Ti-alloys, mild oxidative wear may occur due to slight delamination of tribo-oxide layers and smoothening of matrix surfaces at elevated temperatures. It can be observed that the wear rates in severe wear (abrasive and adhesive wear) are one or two times of magnitude higher than mild oxidative wear [25]. The oxidative wear in the Ti-alloy accelerates when the abrasive effect is exerted by the counter-face material (AISI M2 steel) at minimal sliding speeds. The retained tribo-oxides in the tribo-contact points of the Ti-alloy oxidize to form the tribo-oxide layer, which results in preventing the oxidative wear if it does not form the tribo-oxide layer. This finally results in abrasive wear at the tribo-contact points.

Influence of Frictional Load on Dry Sliding Wear of Ti-Alloys

In Ti-alloys, the subsurface matrix endures noticeable and permanent deformation due to high frictional load and high pressure at the tribo-contact surfaces. This leads to tribo-layer delamination and results in direct metal-to-metal contacts [25]. The lack of wear protection from tribo-oxides on the worn-out surfaces does not strangely adhere to the substrate matrix. Enlargement of a delaminated layer region existed on the worn-out surfaces with load increment. An un-delaminated and a delaminated layer region on the worn-out surfaces resemble a tribo-oxide layer and its delamination, respectively [35].

The load-bearing capacity of the tribo-oxide layers could be measured by the size and density of the plastic flow region at the Ti-alloys subsurface matrix [35]. An increment in applied load had linear progression with the width and depth of the wear tracks and increased the quantity of material layer pile-up on the sides of the wear tracks [23]. With increasing load, the adhesive and oxidative wear increases linearly and the frictional load will hold a significant influence on the abrasive wear mechanism [41].

Influence of Sliding Velocity on Dry Sliding Wear of Ti-Alloys

Higher values of sliding velocities (> 4 m/s) and sliding distance (>3600 m) also produce a remarkable influence on the tribo-oxide layer formation. These layers were compact, thick and contained oxygen that would exhibit lower thermal conductivity and improved load-carrying capability at the worn-out surfaces of the Ti-alloys. Mild wear to severe wear transition happens in the Ti-alloys due to the increase of sliding speed and load [22]. For example, sliding speed at 4 m/s, the value of micro-hardness of the tribo-oxide layer was 770–810 HV with the contact pressure ranging from 0.51 to 2.55 MPa (Fig. **8**).

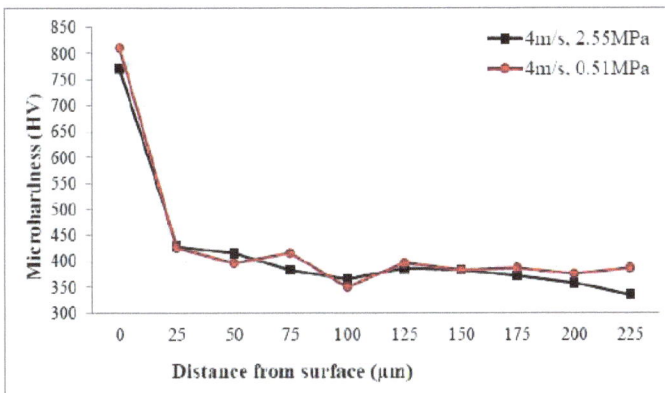

Fig. (8). Micro-hardness of tribo-layers of Ti-alloys.

When the contact pressure increases, the frictional heat at the tribo-contact points also increases and intensifies the metal oxidation. Comparatively, higher micro-hardness of 810 HV was observed at 2.55 MPa, showing the presence of tribo-oxides in the developed compact tribo-oxide layer [33]. Permanent deformation and the retained wear debris in the tribo-contact interfaces will control the wear behavior at low sliding speed and/or contact temperature. When the sliding speed /or contact temperature increases, the contact temperature in the tribo-contact interfaces influences the control of wear behavior [32].

At the critical point of friction strain, the tribo-contacting elements on worn-out surfaces begin to yield permanent deformation and fracture. The number of tribo-oxides increases with increment in sliding speed due to increased tribo-contact temperature, which influences the reduction in friction coefficient. The adhesion between the tribo-contact interface decreases to a significant level of tribo-oxide in the tribo-layer [26]. During lower sliding velocity at room temperature, the development of wear debris and its oxidation occurs due to insufficient energy of oxidation [35]. Increasing the sliding velocity to some extent results in the variation of wear behavior due to the influence of continuous improvement in metallic delamination [20]. The variation in the sliding velocity influences the coefficient of friction (COF) in the following two aspects: (1) Young's modulus decreases as the COF increases due to the increase of the frictional temperature, (2) the ultimate shear strength declines as the COF decreases [42]. The relationship between the COF and sliding velocity differs between wear rate and sliding velocity [44]. The layer thickness, micro-hardness and quantity of tribo-oxides of the tribo-oxide layer in Ti-alloy under various sliding conditions are summarized in Table **4** [32].

Table 4. Properties of the tribo-oxide layer at various sliding conditions

Testing conditions	25°C		200°C		400°C		600°C	
	50N	100-200N	50N	100-200N	50N	100-200N	50N	100-200N
Thickness of tribo-layer (mm)	3-6	20-25	5-8	15-20	15-20	20-30	10-15	10-15
Hardness of tribo-layer (HV)	325	410-440	330	385-440	375	520-570	225	325-475
Hardness of substrate (HV)	280	340	320	340	350	320-360	175	175-210
Hardness difference of tribo-layer and substrate	45	70-100	10	45-100	25	200-210	50	150-265
Amount of tribo-oxides	No	No	Trace	Trace	No	More	Trace	More

Performance of Tribo-Oxide Layers

Wear rate in Ti-alloys can be decided by the performance of tribo-layers, and the performance of tribo-layers can be identified by measuring its micro-hardness. The high value of micro-hardness exhibits the presence of many oxides and compact tribo-oxide layers. Microhardness had a linear progression with the applied load at different temperatures. It can be noted that at a linear distance away from the worn-out surface, the micro-hardness rapidly decreases closer to the tribo-layer and then slightly decreases to the stable value in the Ti-alloy substrate matrix [32, 47].

The outmost micro-hardness might be observed during elevated temperature and peak load because of the presence of many oxides in the tribo-layer and intensified metal oxidation [21]. The micro-hardness distribution in the Ti-alloy subsurface rubbing against various counter-face materials exposed that no significant thermal softening of the element adjacent to the tribo-contact surfaces. The criterion for the defending function of the tribo-oxide layer is the Ti-alloy subsurface matrix holds enough provision for the tribo-layer. During metal-t--metal contact, the provision for the tribo-layer is evaluated by the micro-hardness of the subsurface matrix and softening resistance at high temperatures [25, 48].

Based on the micro-hardness evaluation, the characteristics of two kinds of tribo-oxide layers in the Ti-alloys were listed in Table **5** [38].

Table 5. Characteristics of two kinds of tribo-oxide layers in the Ti-alloys

Layer type	No-oxide layer	Tribo-oxide layer
Appearance	At room temperature	At high temperature or peak sliding speed around room temperature
Morphology	Thin, discontinuous and uncompact	Thick, continuous and compact
Compositions phases	Ti-alloy alone	Ti-alloy with many oxides (TiO_2, Fe_2O_3, Fe_2TiO_5)
Hardness	Almost the same as Ti-alloy	Significantly higher than Ti alloy
Function	No protective role	Protective role

It is noted that during dry sliding of Ti-alloys, the ceramic morphology of the tribo-oxide layer would protect the underneath Ti-alloy subsurface matrix and efficiently reduce the wear rate [38]. The toughness and hardness of the counter-face are greatly influenced by the wear rate of the Ti-alloys [45]. The combination of high strength and ductility offered by the heterogeneous lamellar microstructure will influence the anti-wear property of the Ti-alloys [46, 49].

Mechanisms of Severe-to-Mild Wear Transformation

In dry sliding of Ti-alloys, wear can be categorized into mild and severe wear. According to wear features, mild wear was identified by fine wear debris and flat worn-out surfaces, whereas severe wear was identified by big wear debris and uneven worn-out surfaces. Additionally, based on quantitative analysis, the higher wear rate represents severe wear, and the lower wear rate represents mild wear. The mild and severe wear is simply relative perceptions used to understand the degree of wear damage at the worn-out surfaces [43, 50]. The severe to mild wear behavior of Ti-alloys mainly depends on the frictional temperature at the tribo-contact surfaces.

In a severe wear regime, abrasive and adhesive wear is the principal wear mechanisms, whereas in a mild wear regime, oxidative wear conquered the wear mechanism. Severe-to-mild wear transformation occurs by adding fine tribo-oxide micro particles onto the sliding Ti-alloy surfaces.

The different counter-face materials have a remarkable effect on the wear rate but not on the wear mechanism under various sliding conditions [24]. Similar surface morphologies are observed in the Ti-alloy against different counter-face materials during the rubbing test; it shows that the wear behavior of the Ti-alloy was not affected by the counter-face materials. The volume of material lost during the sliding of Ti-alloys had a linear relationship with the load as well as sliding velocity [37].

CONCLUDING REMARKS

Wear is witnessed to be a complex phenomenon in mechanical engineering and the rate of wear varies if the transition from one wear mechanism to another happens during the wear process. Under dry sliding conditions, metals and their alloys exhibit high adhesion and thus high friction and wear [51, 52]. The above study and analysis provide the knowledge on the effect of the tribo-oxide layer on wear resistance and extend the understanding of rubbing wear of Ti-alloys, resulting in a discovery in the engineering applications of Ti-alloys. It must be emphasized that as far as tribo-oxide appears on the worn-out surfaces of the Ti-alloys, the wear rate would significantly reduce in any sliding conditions and surface treatments for improving the anti-wear property of Ti-alloys can be eliminated considering economic limitations. Furthermore, the research on improving and controlling the development of the tribo-oxide layer of Ti-alloys should be considered as the most important scientific mission with a wider potential engineering application view and scientific merit.

CONSENT FOR PUBLICATION

Not applicable.

CONFLICT OF INTEREST

The author declares no conflict of interest, financial or otherwise.

ACKNOWLEDGEMENTS

First of all, the authors express their sincere gratitude to the editors of this book for providing this wonderful opportunity, and they extend their gratitude to their family for the continuous support provided to accomplish this chapter.

REFERENCES

[1]   Blau PJ. Fifty years of research on the wear of metals. Tribol Int 1997; 30(5): 321-31.
        [http://dx.doi.org/10.1016/S0301-679X(96)00062-X]

[2]   Eyre TS. Wear characteristics of metals. Tribol Int 1976; 9(5): 203-12.
        [http://dx.doi.org/10.1016/0301-679X(76)90077-3]

[3]   Ingole SP. Tribology of Metals and Alloys. InTribology for Scientists and Engineers. New York, NY: Springer 2013; pp. 197-210.
        [http://dx.doi.org/10.1007/978-1-4614-1945-7_6]

[4]   Eyre TS. Wear mechanisms. Powder Metall 1981; 24(2): 57-63.

[http://dx.doi.org/10.1179/pom.1981.24.2.57]

[5]   Randall NX. Experimental Methods in Tribology. InTribology for Scientists and Engineers. New York, NY: Springer 2013; pp. 141-75.
[http://dx.doi.org/10.1007/978-1-4614-1945-7_4]

[6]   Biswas SK, Singh RA. Wear of metals—Influence of some primary material properties. Tribol Lett 2002; 13(3): 203-7.
[http://dx.doi.org/10.1023/A:1020112110162]

[7]   Menezes PL, Nosonovsky M, Kailas SV, Lovell MR. Friction and wear. InTribology for scientists and engineers. New York, NY: Springer 2013; pp. 43-91.
[http://dx.doi.org/10.1007/978-1-4614-1945-7_2]

[8]   Kandeva-Ivanova M, Vencl A, Karastoyanov D. Advanced tribological coatings for Heavy-duty applications: Case studies. Bulgarian Academy of Sciences, Institute of Information and Communication Technologies; Prof Marin Drinov Publishing House of Bulgarian Academy of Sciences. Marin Drinov Publishing House of Bulgarian Academy of Sciences 2016.

[9]   Misra, Ambrish, and Iain Finnie. A review of the abrasive wear of metals. 1982: 94-101.
[http://dx.doi.org/10.1115/1.3225058]

[10]   Richardson RC. The wear of metals by relatively soft abrasives. Wear 1968; 11(4): 245-75.
[http://dx.doi.org/10.1016/0043-1648(68)90175-0]

[11]   Antler M. Sliding wear of metallic contacts. IEEE Transactions on Components, Hybrids, and Manufacturing Technology 1981; 4(1): 15-29.
[http://dx.doi.org/10.1109/TCHMT.1981.1135784]

[12]   Suh NP. An overview of the delamination theory of wear. Wear 1977; 44(1): 1-6.
[http://dx.doi.org/10.1016/0043-1648(77)90081-3]

[13]   Findik F. Latest progress on tribological properties of industrial materials. Mater Des 2014; 57: 218-44.
[http://dx.doi.org/10.1016/j.matdes.2013.12.028]

[14]   Vincent LB, Berthier Y, Dubourg MC, Godet M. Mechanics and materials in fretting. Wear 1992; 153(1): 135-48.
[http://dx.doi.org/10.1016/0043-1648(92)90266-B]

[15]   Ritchie RO. Mechanisms of fatigue-crack propagation in ductile and brittle solids. Int J Fract 1999; 100(1): 55-83.
[http://dx.doi.org/10.1023/A:1018655917051]

[16]   Wang SQ, Wei MX, Wang F, Cui XH, Dong C. Transition of mild wear to severe wear in oxidative wear of H21 steel. Tribol Lett 2008; 32(2): 67-72.
[http://dx.doi.org/10.1007/s11249-008-9361-y]

[17]   Kumar D, Lijesh KP, Deepak KB, Kumar S. Enhancing tribological performance of Ti-6Al-4V using pin on disc setup. InAIP Conference Proceedings 2018 2018 8; 1953(1): 030108. AIP Publishing LLC.

[18]   Król S, Ptacek L, Zalisz Z, Hepner M. Friction and wear properties of titanium and oxidised titanium in dry sliding against hardened C45 steel. J Mater Process Technol 2004; 157: 364-9.
[http://dx.doi.org/10.1016/j.jmatprotec.2004.09.057]

[19]   Abouei V, Saghafian H, Kheirandish S. Dry sliding oxidative wear in plain carbon dual phase steel. J Iron Steel Res Int 2007; 14(4): 43-8.
[http://dx.doi.org/10.1016/S1006-706X(07)60056-9]

[20]   Sahoo R, Jha BB, Sahoo TK, Sahoo D. Effect of microstructural variation on dry sliding wear behavior of Ti-6Al-4V alloy. J Mater Eng Perform 2014; 23(6): 2092-102.
[http://dx.doi.org/10.1007/s11665-014-0987-7]

[21]   Wang L, Zhang QY, Li XX, Cui XH, Wang SQ. Dry Sliding Wear Behavior of Ti-6.5 Al-3.5 Mo-1.5

Zr-0.3 Si Alloy. Metall Mater Trans, A Phys Metall Mater Sci 2014; 45(4): 2284-96.
[http://dx.doi.org/10.1007/s11661-013-2167-z]

[22]   Li XX, Zhou Y, Li YX, Ji XL, Wang SQ. Dry sliding wear characteristics of Ti-6.5 Al-3.5 Mo-1.5 Zr-0.3 Si alloy at various sliding speeds. Metall Mater Trans, A Phys Metall Mater Sci 2015; 46(9): 4360-8.
[http://dx.doi.org/10.1007/s11661-015-3019-9]

[23]   Farokhzadeh K, Edrisy A. Transition between mild and severe wear in titanium alloys. Tribol Int 2016; 94: 98-111.
[http://dx.doi.org/10.1016/j.triboint.2015.08.020]

[24]   Wang L, Li XX, Zhou Y, Zhang QY, Chen KM, Wang SQ. Relations of counterface materials with stability of tribo-oxide layer and wear behavior of Ti–6.5 Al–3.5 Mo–1.5 Zr–0.3 Si alloy. Tribol Int 2015; 91: 246-57.
[http://dx.doi.org/10.1016/j.triboint.2015.01.028]

[25]   Chen KM, Zhang QY, Li XX, Wang L, Cui XH, Wang SQ. Comparative study of wear behaviors of a selected titanium alloy and AISI H13 steel as a function of temperature and load. Tribol Trans 2014; 57(5): 838-45.
[http://dx.doi.org/10.1080/10402004.2014.916372]

[26]   Straffelini G, Molinari A. Mild sliding wear of Fe–0.2% C, Ti–6% Al–4% V and Al-7072: a comparative study. Tribol Lett 2011; 41(1): 227-38.
[http://dx.doi.org/10.1007/s11249-010-9705-2]

[27]   Zhong H, Dai LY, Yue Y, *et al.* Friction and wear behavior of annealed Ti-20Zr-6.5 Al-4V alloy sliding against 440C steel in vacuum. Tribol Int 2017; 109: 571-7.
[http://dx.doi.org/10.1016/j.triboint.2017.01.040]

[28]   Zhou Y, Wang S, Chen W, *et al.* Comparative research on the effect of an oxide coating and a tribo-oxide layer on dry sliding wear of Ti–6Al–4V alloy. Proc Inst Mech Eng, Part J J Eng Tribol 2018; 232(12): 1569-80.
[http://dx.doi.org/10.1177/1350650118757329]

[29]   Wang L, Zhang QY, Li XX, Cui XH, Wang SQ. Severe-to-mild wear transition of titanium alloys as a function of temperature. Tribol Lett 2014; 53(3): 511-20.
[http://dx.doi.org/10.1007/s11249-013-0289-5]

[30]   Cui XH, Mao YS, Wei MX, Wang SQ. Wear characteristics of Ti-6Al-4V alloy at 20–400° C. Tribol Trans 2012; 55(2): 185-90.
[http://dx.doi.org/10.1080/10402004.2011.647387]

[31]   Chen KM, Zhou Y, Li XX, Zhang QY, Wang L, Wang SQ. Investigation on wear characteristics of a titanium alloy/steel tribo-pair. Materials & Design (1980-2015) 2015; 65: 65-73.

[32]   Mao YS, Wang L, Chen KM, Wang SQ, Cui XH. Tribo-layer and its role in dry sliding wear of Ti–6Al–4V alloy. Wear 2013; 297(1-2): 1032-9.
[http://dx.doi.org/10.1016/j.wear.2012.11.063]

[33]   Li XX, Zhou Y, Ji XL, Li YX, Wang SQ. Effects of sliding velocity on tribo-oxides and wear behavior of Ti–6Al–4V alloy. Tribol Int 2015; 91: 228-34.
[http://dx.doi.org/10.1016/j.triboint.2015.02.009]

[34]   Heilig S, Ramezani M, Neitzert T, Liewald M. Tribological performance of duplex-annealed Ti-6A--2Sn-4Zr-2Mo titanium alloy at elevated temperatures under Dry sliding condition. J Mater Eng Perform 2018; 27(4): 2003-9.
[http://dx.doi.org/10.1007/s11665-018-3264-3]

[35]   Zhang QY, Zhou Y, Li XX, Wang L, Cui XH, Wang SQ. Accelerated formation of tribo-oxide layer and its effect on sliding wear of a titanium alloy. Tribol Lett 2016; 63(1): 2.
[http://dx.doi.org/10.1007/s11249-016-0694-7]

[36]   Zhang QY, Zhou Y, Wang L, Cui XH, Wang SQ. Investigation on tribo-layers and their function of a titanium alloy during dry sliding. Tribol Int 2016; 94: 541-9.
[http://dx.doi.org/10.1016/j.triboint.2015.10.018]

[37]   Hadke S, Khatirkar RK, Shekhawat SK, Jain S, Sapate SG. Microstructure evolution and abrasive wear behavior of Ti-6Al-4V alloy. J Mater Eng Perform 2015; 24(10): 3969-81.
[http://dx.doi.org/10.1007/s11665-015-1667-y]

[38]   Li XX, Zhang QY, Zhou Y, Liu JQ, Chen KM, Wang SQ. Mild and severe wear of titanium alloys. Tribol Lett 2016; 61(2): 14.
[http://dx.doi.org/10.1007/s11249-015-0637-8]

[39]   Kumar D, Deepak KB, Muzakkir SM, Wani MF, Lijesh KP. Enhancing tribological performance of Ti-6Al-4V by sliding process. Tribology-Materials. Surf Interfaces 2018; 12(3): 137-43.
[http://dx.doi.org/10.1080/17515831.2018.1482676]

[40]   Jeyaprakash N, Yang CH. Microstructure and Wear Behaviour of SS420 Micron Layers on Ti–6Al–4V Substrate Using Laser Cladding Process. Trans Indian Inst Met 2020; 12: 1-7.
[http://dx.doi.org/10.1007/s12666-020-01927-7]

[41]   Heilig S, Ramezani M, Neitzert T, Liewald M. Investigation of Friction and Wear Properties of Duplex-Annealed Ti–6Al–2Sn–4Zr–2Mo Against Hardened AISI E52100 at Linear Reciprocating Motion. Trans Indian Inst Met 2018; 71(5): 1257-64.
[http://dx.doi.org/10.1007/s12666-017-1262-z]

[42]   Zhong H, Dai LY, Yang YJ, *et al.* Vacuum Tribological Properties of Ti-20Zr-6.5 Al-4V Alloy as Influenced by Sliding Velocities. Metall Mater Trans, A Phys Metall Mater Sci 2017; 48(11): 5678-87.
[http://dx.doi.org/10.1007/s11661-017-4301-9]

[43]   Zhang Q, Ding H, Zhou G, *et al.* Dry Sliding Wear Behavior of a Selected Titanium Alloy Against Counterface Steel of Different Hardness Levels. Metall Mater Trans, A Phys Metall Mater Sci 2019; 50(1): 220-33.
[http://dx.doi.org/10.1007/s11661-018-4993-5]

[44]   Yang LQ, Zhong H, Lv G, *et al.* Dry sliding behavior of a TiZr-based alloy under air and vacuum conditions. J Mater Eng Perform 2019; 28(6): 3402-12.
[http://dx.doi.org/10.1007/s11665-019-04100-4]

[45]   Liang SX, Yin LX, Zheng LY, *et al.* Tribological behavior and wear mechanism of TZ20 titanium alloy after various treatments. J Mater Eng Perform 2018; 27(9): 4645-54.
[http://dx.doi.org/10.1007/s11665-018-3570-9]

[46]   Zhang GS, Guo DF, Li M, *et al.* Tailoring microstructure and tribological properties of cold deformed TiZrAlV alloy by thermal treatment. Acta Metall Engl Lett 2017; 30(5): 493-8.
[http://dx.doi.org/10.1007/s40195-017-0558-7]

[47]   Jeyaprakash N, Duraiselvam M, Aditya SV. Numerical modeling of WC-12% Co laser alloyed cast iron in high temperature sliding wear condition using response surface methodology. Surf Rev Lett 2018; 25(07): 1950009.
[http://dx.doi.org/10.1142/S0218625X19500094]

[48]   Jeyaprakash N, Duraiselvam M, Raju R. Modelling of Cr3C2-25% NiCr laser alloyed cast iron in high temperature sliding wear condition using response surface methodology. Arch Metall Mater 2018; 63.

[49]   Jeyaprakash N, Yang CH, Tseng SP. Wear Tribo-performances of laser cladding Colmonoy-6 and Stellite-6 Micron layers on stainless steel 304 using Yb: YAG disk laser. Met Mater Int 2019; 13: 1-4.

[50]   Jeyaprakash N, Yang CH, Tseng SP. Characterization and tribological evaluation of NiCrMoNb and NiCrBSiC laser cladding on near-α titanium alloy. Int J Adv Manuf Technol 2020; 106(5): 2347-61.
[http://dx.doi.org/10.1007/s00170-019-04755-2]

[51]   Jeyaprakash N, Yang CH. Improvement of tribo-mechanical properties of directionally solidified CM-

247 LC nickel-based super alloy through laser material processing. Int J Adv Manuf Technol 2020; 106(11): 4805-14.
[http://dx.doi.org/10.1007/s00170-020-04931-9]

[52]   Jeyaprakash N, Yang CH, Duraiselvam M, Prabu G. Microstructure and tribological evolution during laser alloying WC-12% Co and Cr3C2− 25% NiCr powders on nodular iron surface. Results Phys 2019; 12: 1610-20.
[http://dx.doi.org/10.1016/j.rinp.2019.01.069]

CHAPTER 3

Hardfacing Alloy Powders

S. Venukumar[1], T. Vijaya Babu[1] and K. Tejonadha Babu[2, *]

[1] *Department of Mechanical Engineering, Vardhaman College of Engineering, Hyderabad-501218, Telangana, India*

[2] *Department of Mechanical Engineering, Kallam Haranadhareddy Institute of Technology, Guntur-522019, Andhra Pradesh, India*

Abstract: Hardfacing alloying is one of the inevitable methods to protect the components which are used in heavy-duty machinery. Specifically, this process is performed to improve the tribological properties of industrially important materials, such as AISI 304 stainless steel (SS) bushes in nuclear power plants, SS 316 control valves in coal power plants, polycrystalline martensitic steel valve seats in steam turbines, aluminum-based body sheet materials in the automobile, Co-based, Fe-based and Ni-based alloys. These materials have unique advantages to use for their specific application. However, these materials are preferred to use for hard-facing as those alloys have poor tribological properties at the different working conditions as they are prone to abrasive, adhesive wear and corrosion. Tribological properties, such as hardness and wear resistance of the alloyed material, mainly depend on the proper selection of hardfacing alloy powders, which have equal importance in the selection of process. Colmonoy-5, Colmonoy-6, Stellite-6, Inconel-625, H13 Steel, stainless steel 420, stainless steel 304, Ti-64, IN-718, AlSiMg, Scalmalloy, SS316L and CuCrZr are some of the important hardfacing alloy powders to enhance the tribological properties. Laser surface treatments, Gas Tungsten Arc-based alloying and Plasma Transferred Arc are the major hardfacing processes. Laser surface treatments are one of the most effective surface alloying processes for enhancing materials properties due to their unique advantages over other processes, such as flexible operation, high energy density and chemically clean operation. Moreover, it produces good metallurgical bonding of the alloyed region with the substrate material, minimizes the heat-affected zone (HAZ) and forms finer microstructure. This chapter provides a detailed understanding of various research works carried out on hardfacing alloying by laser-treated processes on different substrates. The outcome of this chapter would be beneficial for the present and future research based on laser-based hardfacing and also for many hardfacing industries.

Keywords: AlSiMg, Coefficient of friction, Colmonoy powder, CuCrZr, Inconel powder, Stellite powder, Scalmalloy, Tribological property, Wear resistance.

---

[*] **Correspondence author K. Tejonadha Babu**: Department of Mechanical Engineering, Kallam Haranadhareddy Institute of Technology, Guntur-522019, India; E-mail: tejonadhababu@gmail.com

Jeyaprakash Natarajan and Che-Hua Yang (Eds.)
All rights reserved-© 2021 Bentham Science Publishers

INTRODUCTION

The main purpose of using the hardfacing alloy powders is to enhance the hardness, wear-resistance and reduce the coefficient of friction. Various hardfacing alloy powders are available in the market. After the hardfacing process, a protective alloyed region is formed on the substrate that resists different types of wear. Each powder has a specific application due to its tribological behavior. Will the use of hardfacing alloy powders suppress all or only a few types of wear? The practicing industrial engineer wants to answer this and some other questions. This chapter discusses the fundamental phase formation during the alloying/cladding processes, the protective coating, and the application of industrially important hardfacing alloy powders.

TRIBOLOGICAL PROPERTIES OF HARDFACING ALLOY POWDERS

Wear plays an important role when the component makes contact with the mating parts during the operation time. There will be a loss of material due to the wear of the components that lead to the reduction in performance and also increase the downtime of equipment. The coating on the component by the laser surface alloying process would enhance hardfacing process on the new as well as the worn-out components. It tends to extend the service life of the components and avoid the need for replacement. Hardfacing of alloy powders through laser-based surface modification techniques has become an effective method for many industrial applications, such as chemical, marine, motorized equipment, energy production, agriculture, heavy automotive, petrochemical, glass mold, paper, plastic extrusion and steel manufacture. These applications need resistance to abrasive wear, achieving wear, galling, fretting, heat, cavitation erosion, impingement erosion, corrosion and impact.

The sliding contact between the two mating parts causes the adhesive wear. The contact among the asperities, peaks and valleys of the matting parts results in severe wear and worn-outs. Adhesive wear arises due to the sliding action between the two metallic components. The severity of adhesive wear would seize the movement of the matting parts. Wear debris would act as the third body abrasive element that induces scratches on the matting surfaces. Similarly, material loss occuring due to the moving particles is referred to as erosive wear. Air particles collapsing on the metal surface leads to the erosion of the surface due to the cavitation during turbulent fluid flow. This turbulent flow caused a shock wave on the metal surface. Fritting wear was induced by the synergic effect of abrasive and adhesive wear due to small vibration at the matting surface.

Hardfacing alloy powders are alloyed on the surface of the different components by laser surface treatment, HVOF, plasma spray and Plasma Transferred Arc process. These powders enhance the surface properties, such as hardness, wear and corrosion resistance. These powders are produced by different methods, such as vacuum inert gas atomized and cooled in nitrogen or argon inert gas. This process produces spherical shape particles with a homogeneous chemical composition. These powders are used to enhance the service life of components. This chapter briefly discusses these phase formations from the powder, which is required for the above-specified application.

TYPES OF HARD FACING ALLOY POWDERS

Nickel Based Alloy Powders

The Ni-based Colmonoy family powder offers higher wear protection and retains the microhardness up to the temperature of 600°C. It is applied on the surface of the components for enhancing the hardness and wear resistance by laser surface alloying, thermal spraying, or a similar type of process. Hardfacing alloy powders can be used on the new components or to repair worn-out components. Acceptable forms of material for hardfacing powders are solid rods, wires, tubular rods/wires. Compared to other forms of raw material, powders have a greater influence on the tribological property due to the uniform distribution of alloying elements. Water and gas atomization are widely used techniques for producing the hardfacing alloy powders. The particle size of powders used for hardfacing processes is within the range of 10 to 150 mm. Powder shape is an important parameter that affects the flowability of powder in any of the hardfacing operations. The flow of spherical powders is better than that of irregular ones when using gravity powder feed devices. Gas atomization techniques produce spherical powders, whereas water atomized techniques produce both spherical and irregular powders depending on cooling rate and alloy type.

Colmonoy-5 Powder

Nickel-based hardfacing alloys are grouped as boride-containing alloys, carbide-containing alloys, and laves phase-containing alloys. Colmonoy 5 alloy powder comes under the category of boride-containing alloys. The chemical composition of the Colmonoy-5 alloy powder is given in Table **1**. Scanning Electron Micrograph (SEM) image of Colmonoy-5 powder is shown in Fig. (**1**). A uniform spherical shape of Colmonoy-5 powder is observed. The Colmonoy-5 alloy powder contains the complex structure of Ni-B-Cr-Si-C-Fe, which possesses hexagonal close-packed Cr_7C_3 structure and face-centered cubic $Cr_{23}C_6$ structure. It has self-fluxing properties due to the presence of silicon and boron. The powder has 10 - 15 wt.% of C-B-Cr precipitates that are dispersed in the Ni-Cr-B-Fe solid solution matrix. Chromium carbide (Cr_7C_3), chromium carboboride ($Cr_{27}BC_4$) and chromium boride phases are formed in the range of 60, 20, 10 wt.%., respectively. Ni-Fe-B phases are also formed, which are not in the pure form.

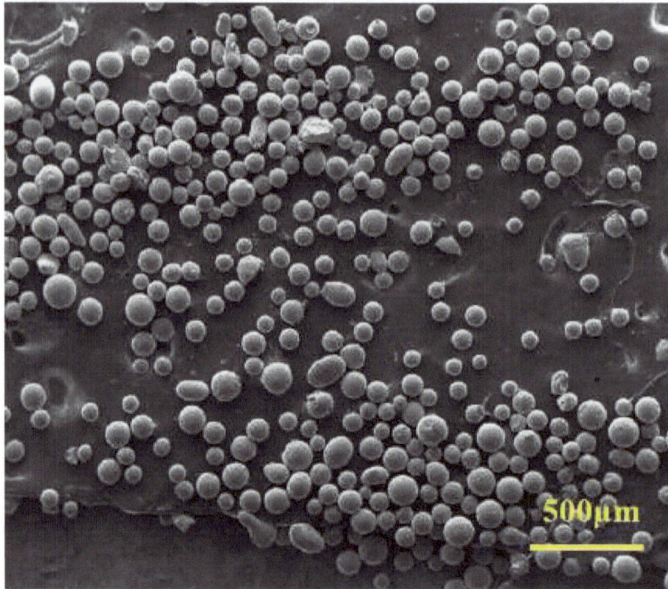

Fig. (1). SEM image of Colmonoy-5 powder.

Table 1. Composition of Colmonoy-5 powder

Elements	Ni	Cr	Fe	Si	B	C
Wt (%)	Bal.	14.3	4.9	4.8	1.6	0.74

The laser power, scanning velocity of the laser beam, and the beam diameter of the beam are considered while alloying the Colmonoy-5 powder onto the substrate. The optimized parameters are achieved through a continuous comparison between the properties of the processing parameters and the crack-free Colmonoy-5 coatings. The laser alloyed region of the coating possesses a thickness of up to 1 mm thick with a dendritic and compact homogeneous structure. The laser alloyed Colmonoy-5 region has the distribution of the elements of the matrix and precipitates at the grain boundary. Carbon, boron and chromium elements segregate as carboborides and carbides in the laser alloyed Colmonoy-5 coating. The resultant coating shows nickel-boron based intermetallic components, such as Ni_3B_4, Ni_4B_3 and Ni_3B. It also contains different hard phases, such as Cr_5B, Cr_2B and CrB and other boride compounds. The alloy hardness mainly depends on the boron content. Chromium boride has a higher hardness of 4600 HV, which is very brittle. Sharp variation of microhardness profiles would appear at the interface between the substrate and the coating due to the presence of hard particles at the coated region. Chromium increases the microhardness of the laser alloyed region by forming very hard precipitates. Moreover, it improves the resistance to oxidation and corrosion. Boron contributes to forming different hard phases. Silicon increases the self-fluxing properties of the alloyed region and reduces the coefficient of friction. Carbides are formed in the alloyed region due to the presence of carbon that improves the hardness levels as well as the wear resistance of the laser alloyed region.

Corchia M *et al.* [1] have coated Colmonoy-5 hardfacing alloy powder on austenitic stainless-steel substrate that is used in nuclear plants. The alloying process was performed by using laser surface modification techniques. Their result revealed that the wear resistance was increased due to the presence of carbides and borides hard phases in the laser alloyed region. Gnanasekaran S *et al*. [2] have deposited the Colmonoy-5 Ni-based powder on the 316LN austenitic stainless-steel substrate and studied the tribological properties. The laser surface alloying process was used to repair or modify the surface of the substrate. The microstructure and tribo-properties of the alloyed region were studied by varying the powder feed rate. It was identified that the hardness of the laser alloyed region increased to 800 HV with the increase of powder feed rate. The microstructure of the laser alloyed hardfacing consists of chromium-rich carbides, nickel-rich borides and γ-nickel. It contributed to increasing the hardness at a higher powder feed rate.

Colmonoy-6 Powder

Colmonoy-6 is used in nuclear power plants due to the property of low induced radioactivity. Colmonoy-6 bushes are used in the local hardfacing place, where the locations and geometries are not easily accessible and complicated, respectively. Conventionally, it is fabricated by the casting technology and gas tungsten arc alloying process. These multi-stage processes consume more machining time. Moreover, high manufacturing setup costs and limited manufacturing volume result in an unaffordable cost of the bushes. To meet the quality and economic constraints, these Colmonoy-6 bushes are fabricated by laser surface modification techniques. It deposits the hardfacing alloy powder on austenitic steel rods, which would reduce the secondary machining cost and loss of Colmonoy-6 powder. The composition of Colmonoy-6 is shown in Table **2**. SEM image of Colmonoy-6 powder is shown in Fig. (**2**). The Colmonoy-6 powder has a higher content of boride compared to Colmonoy-5, which increased the hardness of the Colmonoy-6 component.

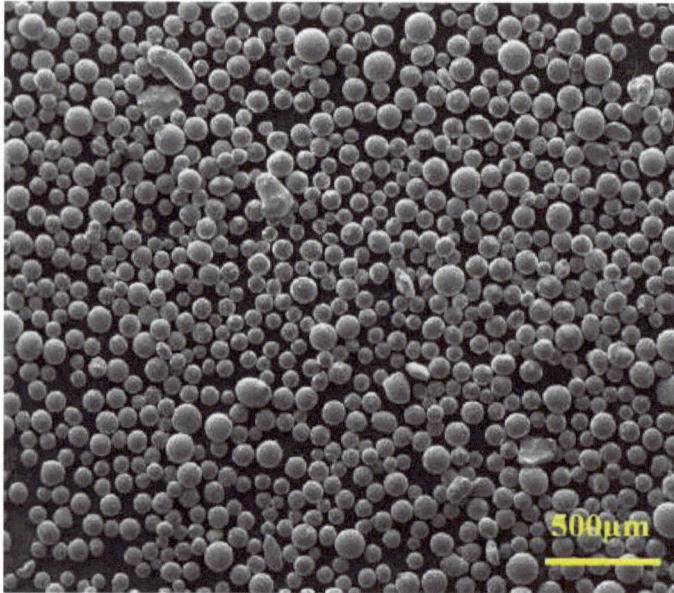

Fig. (2). SEM image of Colmonoy-6 powder.

Table 2. Composition of Colmonoy-6 powder

Elements	Ni	Cr	Fe	Si	B	C
Wt (%)	Bal.	14.3	4.0	4.3	3.0	0.74

The laser alloyed region of the Colmonoy-6 bush has a uniform dendrite structure that is grown in the build-up direction of the alloyed region. Dendrite phases and inter-dendrite structures are formed in the alloyed region. The dendrite structure is formed with the content of g- nickel phase, Si, Fe and Cr. Eutectic phases are formed in between the g- nickel phase dendrites. Borides have appeared as dark components, whereas g- nickel phase appeared as a lighter component in the microstructure analysis. However, a slight variation in the chemical composition occurred in the direction of the substrate to the laser alloyed region due to the difference in dilution rate of hardfacing alloying elements. Borides and carbides compounds like CrB have existed in the laser alloyed layer that increased the hardness of the alloyed region. Zhang H *et al.* [3] studied the tribological property of stainless steels, which are used in the nuclear power plant due to better corrosion resistance. However, the wear resistance of stainless steel is relatively low. To improve the surface property of stainless steel, the Colmonoy-6 powder was coated on AISI316L austenitic steel by CO_2 laser. The result revealed that the wear resistance of the laser cladding region was 53 times higher than that of the substrate material due to the formation of Ni-rich austenitic elements. Jeyaprakash

N *et al.* [4] studied the wear behavior of laser cladding Colmonoy-6 on stainless steel 304. The laser cladding region was examined to study their mechanical, metallurgical and tribological properties. The outcomes indicated that the coatings possess higher hardness and wear resistance than the substrate due to the formation of a dendrite structure. Moreover, abrasive and adhesive wear mechanisms have majorly occurred during the tribological test. The wear resistance of the Colmonoy-6 coating was 49 times higher than that of the substrate sample. It shows that laser cladding plays an important role in the wear resistance of the substrate.

Paul C P *et al.* [5] have fabricated Colmonoy-6 components using laser rapid manufacturing (LRM) techniques. LRM is one of the 3D printing techniques, where the near-net shaping of the components can be manufactured by depositing the Colmonoy-6 powder in the layer-by-layer form from the CAD model. High power CO_2 continuous wave (CW) laser with the co-axial powder feeding system was used. The effect of laser processing parameters for manufacturing the Colmonoy -6 bushes has been studied. The phase analysis of laser alloyed region of Colmonoy-6 bushes was examined and found that the formation of different carbide and boride phases, such as $CrTC_3/Ni_4B_3$, Cr_6B_3/Ni_3B, $Cr_6B_3/Ni_2B/CrB$, $Ni/Ni_4B_1/Cr_7C$, $CrTC_3/Cr_2B/Ni_4B_3/NiB$, Ni_2B/Cr_2B, Ni. As a result of the formation of hard phases, both the hardness and wear resistance of the LRM system are higher than that of the gas tungsten arc welding techniques.

Inconel-625 Powder

Power output, cycle efficiency, and maintenance cost of steam turbine depend on the provision of admission valves. The efficient operation of valves satisfies the continuous demand for electrical energy. Friction loads and high impact on the admission valves are prominent issues that occur during the operation condition. These control valves seal the steam flow in 200 millisecond, which affects the performance of the valves. High load acts on the sealed surface of the valve, which is unavoidable. In a steam power plant turbine, the cast or forged materials are reliable to protect the surface against the failure of the component due to severe wear. Hardfacing cladding is one of the functional coatings to protect the cones, valve seats and other positions that are mainly exposed to friction loads.

Inconel-625 powder is one of the hardfacing alloy powders to clad on the sealed surface of the valve. The chemical composition of Inconel-625 is shown in Table **3**. SEM image of Inconel-625 powder is shown in Fig. (**3**). The Inconel-625 powder has a higher amount of Cr compared to Colmonoy-6. The Fe elements in the cladding region play an important role in enhancing the hardness of the coating. The cladding elements, such as g- Ni FCC phase, Mo and Nb are the strengthening elements that increase the wear resistance of the Inconel 625 cladding sample.

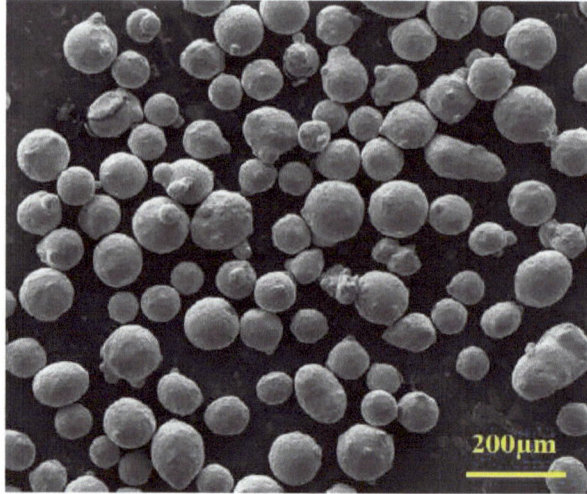

Fig. (3). SEM image of Inconel-625 powder.

Table 3. Composition of Inconel-625 powder

Elements	Ni	Cr	Mo	Fe	Co	Al	Si	Mn
Wt (%)	Bal.	22.8	10.0	5.0	1.0	0.5	0.5	0.5

Feng *et al.* [6] fabricated the Inconel 625 coatings by the shielded metal arc and laser cladding methods. The tribological properties of the cladding region at different temperatures were studied. The results revealed that the laser cladding of the Inconel 625 coating has a finer microstructure. Mo, Fe and Nb elements are segregated in the shielded metal arc alloying process. The laser cladding region has a higher hardness in the coated region, which is much higher than that of the iron dilution zone and the formation of finer microstructure. The laser cladding region has higher wear resistance due to the lower dilution of the Fe element and shows higher hardness. The cladding coating is preferred due to better performance at both room and elevated temperatures. Anil PM *et al.* [7] studied the wear behavior of the Inconel 625 by varying the process parameters, such as the load, temperature and coating thickness. The sliding wear test was performed, and the wear properties were investigated. It was found that the wear depth of 2.88 μm has occurred on the cladding region of the Inconel 625 sample, which is less than that of the substrate due to the presence of the hardfacing Inconel 625 alloy elements.

Inconel 718 Powder

Control rod in the drive system, steam power generators in the nuclear power plants, thrust bearings in the compressors, thrust propulsion for hypersonic aircraft rockets and missiles, shafts with turbine discs, bearings, pressure lines, hot working tools, rings, dies fasteners, and bolts in the turbine of the aero-engine experience higher wear during its operating condition. Wear occurs in the form of adhesion, erosion, delamination, abrasion and melting. To meet the industry's requirements, such as high load-bearing capacities, greater precision, longer service lifespan and compactness, hardfacing alloying powders are used to reduce the degradations of machine elements. Because degradation under severe wear conditions affects the dimensions and surface finish of the machine components, it results in vibrations, clattering, reduces the service life, loss of efficiency, and serves as the source of catastrophic failure. Particularly, wear induces several complications to the mating components by degrading the properties of materials.

The demand for retaining the strength capacity for these types of applications makes the Ni-based Inconel 718 hardfacing alloy a natural choice. The chemical composition of Inconel-718 powder is tabulated in Table **4**. SEM image of Inconel-718 powder is shown in Fig. (**4**). It is an age hardenable hardfacing alloy that retains strength up to 700 °C. The strength of the cladding region is retained by the inducement of different hard phases such as γ"-Ni_3 (Al, Nb, Ti)) and γ'-Ni (Ti, Al) in the γ-phase FCC matrix. Calvin Samue *et al*. [8] studied the dry sliding wear behavior of Inconel 718 fused by laser powder bed techniques. The tribological behaviour of the Inconel 718 alloy was examined, and it was found that the material loss is due to the increase in the wear and friction coefficient with an increase in temperature.

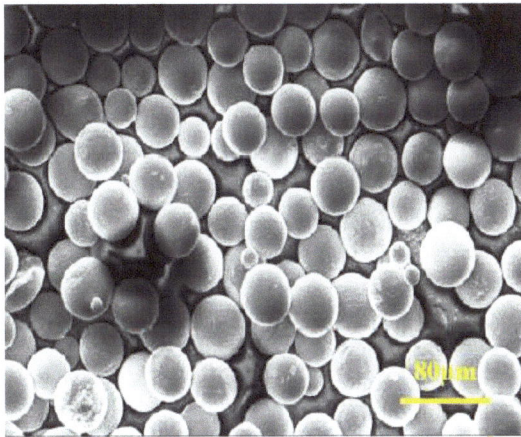

Fig. (4). SEM image of Inconel-718 powder.

Table 4. Composition of Inconel-718 powder

Elements	Ni	Cr	Fe	Mo	Ni	Mn
Wt (%)	Bal.	19.0	17.0	3.0	5.0	0.35

The abrasion, delamination and oxidation wear have predominantly occurred during the wear test. Cr elements in the hardfacing powder formed as Cr_2O_3 and $Cr_{23}C_6$, strong carbide phases in the cladding region. Ni element has a higher content than the Cr elements, which promotes the strengthening of the cladding region due to carbide and solid solution formation. This strengthening process enhanced the wear resistance of the cladding region than the substrate region. Zhanyong Zhao *et al.* [9] studied the wear and friction behavior of Inconel 718 hardfacing alloy fabricated by selective laser melting (SLM). Their results showed that the friction and wear of the alloy had been improved. The leaves phase of (Co, Fe, Ni)2(Mo, Ti, Nb) was formed during the SLM process. It dissolved in Ni_3Nb γ'' phase and precipitated as Ni_3Nb δ phase due to the double aging process. Moreover, γ'' and γ' (Ni3(Al, Ti)) phases were homogeneously formed in the matrix. It improved the microhardness and the wear resistance of hardfacing SLM IN718 material.

Cobalt Based Alloy Powder (Stellite 6)

Co-based hardfacing alloys are used to improve wear, oxidation and corrosion resistance. Among the Co-based hardfacing alloy, the Co-Cr-W alloy, also called Stellite-6, is mostly used in industrial applications. The chemical composition of Stellite 6 powder is shown in Table **5**. SEM image of Stellite 6 powder is shown in Fig. (**5**). In surface modification techniques, Stellite-6 alloys are extensively used to protect various machine parts against different types of wear, namely abrasion, oxidation, erosion, corrosion and cavitation. Stellite-6 hardfacing coatings can be applied in the oil and gas industry. The formation of $M_{23}C_6$ and M_7C_3 carbides contribute to increasing the strength of the cladding region. Moreover, molybdenum (Mo) and tungsten (W) improve the hardness and strength through the precipitation hardening process, which possesses both high density and ductility. The laser cladding process can overcome the formation of insufficient coating and delamination. It is able to produce strong metallurgical bonding between the substrate and coating. As a result, the cladding region possesses minimal distortion, negligible heat affected zone, good process flexibility and low dilution. Stellite 6 coating can be applied to protect the components, such as gas turbine blades, engine valve seats, control valve seats, exhaust pipes, valves in alumina refineries and mold for plastic injection.

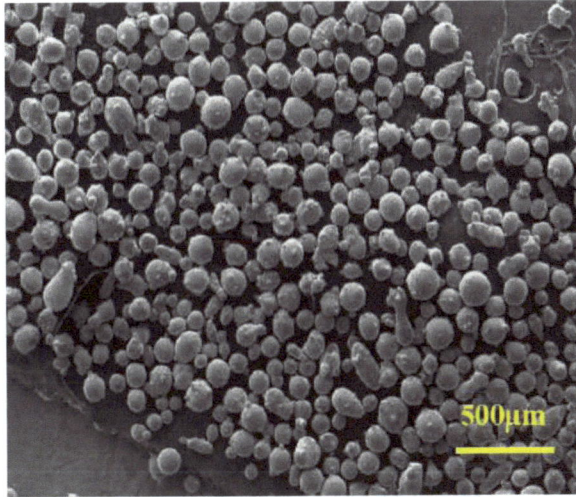

Fig. (5). SEM image of Stellite-6 powder.

Table 5. Composition of Stellite 6 powder

Elements	Co	Cr	W	C	Si	Fe	Ni	Mo
Wt (%)	Bal.	32.0	11.0	3.0	1.2	1.0	1.0	1.0

Bo Li *et al.* [10] deposited the Stellite-6 powder on the substrate using laser energy aid with cold spray techniques. The microstructure and wear-resistant properties of the laser cladding stellite-6 coatings were investigated. The results revealed that the coated region has higher wear resistance due to the formation of fine microstructure and hard phases of carbide and tungsten. The wear behavior of laser cladding Stellite-6 hardfacing alloys was studied with AISI 4140 and AISI 4340 steel by using the pin-on-disc wear test machine. Load and sliding velocity were varied in the range of 9.8 N to 156.8 N and 1 m/s to 4 m/s, respectively. CrO, W_3O, Cr_2O_3, Co_2O_3 and Cr_5O_{12} oxide films were formed on the worn-out surface of the alloyed region, which was tougher than the oxide films on the steel surface. As a result, the laser cladding surface possessed higher wear resistance than the substrate surface.

Steel Based Alloy Powders

H13. Steel

H13 steel powder has excellent properties, such as high resistance to crack formation due to thermal fatigue, high toughness and high stability in heat treatments. The composition of H13 steel powder is shown in Table **6**. SEM image of H13 steel powder is shown in Fig. (**6**). Due to the high strength carbide forming elements, such as Mo, V and Cr, H13 steel has high hardenability. Owing to its properties, H13 steel is the inevitable material used in a different industry for tooling in heavy-duty work applications, such as extrusion, forging, and die casting. H13 steel powder can be cladded on the material where the failures of the component occurr due to wear, corrosion, cracking, thermal fatigue, and pitting. The failure occurs at a small portion of the part, which leads to replacing the entire part. Laser cladding provides a good metallurgical bonding with the steel substrate that guarantees the high wear resistance of the substrate material. Different carbides, such as $Cr_{23}C_6$, Cr_7C_3, VC and Mo_2C, are formed in the cladding region. Moreover, the cladding region has a major constituent of martensite microstructure along with different forms of carbides. Chromium carbides are the most dominant component and precipitate between martensitic laths, whereas Mo and V containing carbides are found between the inter-dendritic and martensitic laths spaces.

Fig. (6). SEM image of H13 steel powder.

Table 6. Composition of H13 steel powder

Elements	Fe	Cr	Mo	Si	V	Mn	C
Wt (%)	Bal.	5.0	1.7	0.8	1.0	0.4	0.4

Gururaj Telasang *et al.* [11] studied the microstructure and tribological properties of laser cladding with AISI H13 steel powder. Laser energy of 133 J/mm^2 was coupled with a co-axial powder feeder, which had the capacity of feeding at the rate of 13.3×10^{-3} g/mm^2. The laser cladding region has 640 VHN hardness, which is higher than the substrate surface. Wear resistance of the laser surface clad had superior performance when compared to that of conventional hardened AISI H13 steel. Telasang G *et al*. [12] studied the microstructure of laser cladding AISI H13 steel powder by using a 6kW fiber laser. The microhardness of the clad zone was evaluated. The maximum improvement in microhardness was 45% higher than that of hardened substrates due to the presence of various forms of carbides.

Steel 304

The composition of steel 304 powder is shown in Table 7. SEM image of steel 304 powder is represented in Fig. (7). The steel 304 laser-based cladding region contains austenitic with a small amount of g-ferrite phase.

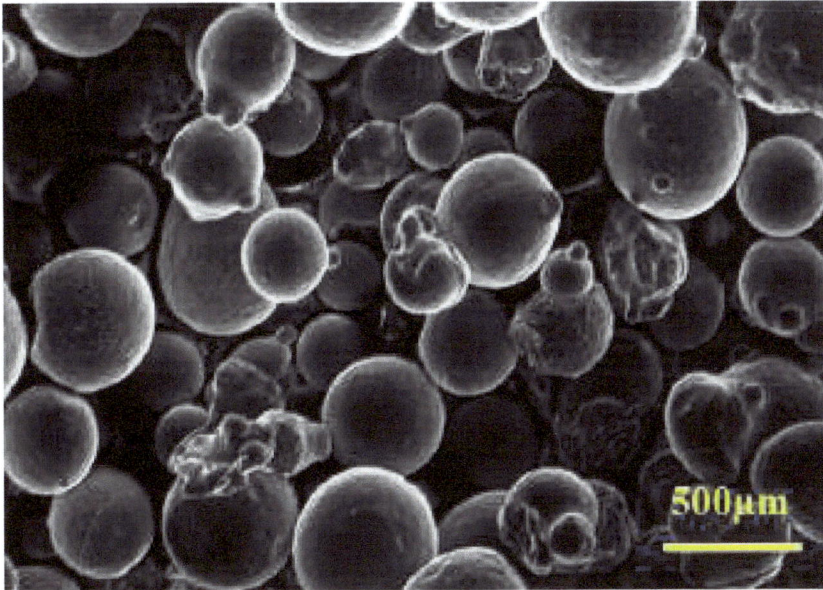

Fig. (7). SEM image of steel 304 powder.

Table 7. Composition of steel 304 powder

Elements	Fe	Cr	Ni	Si	Mn
Wt (%)	Bal.	19.7	11.0	1.1	0.4

Uniform thickness of the clad layer can be obtained through the laser surface modification process. Moreover, a crack and porous free cladding region would be formed. The d-ferrite phase forms at the grain boundaries of the cladding region, which will be appeared as a dark region in the microstructural graph. In addition to the d-ferrite phase, some d-ferrite phase transforms into the to g-ferrite phase, which results in the formation of austenitic structure. d-ferrite phase favors the prevention of the hot cracking of the 304 stainless steel. The ferrite phase formation can be promoted by the high content of the Mo element and the low content of the Ni element. Fouquet F *et al.* [13] studied the coating of AISI 304 grade stainless steel. The coating was performed through laser surface modification techniques. The result revealed that the coating has better metallurgical bonding with the substrate material, which enhanced the tribological property due to the formation of hard phases such as martensite.

Steel 316L

Austenitic 316L stainless steel is one of the engineering alloys for industrial applications, such as automotive and petrochemicals because of its high weldability, good corrosion resistance and high ductility. It is also used in the biomedical industry due to its high deformation energy, biocompatibility and excellent combination of ductility and strength. However, its use in structural industries is limited owing to its low yield strength. The composition of steel 316L powder is shown in Table **8**. SEM image of steel 316L powder is shown in Fig. (**8**). A uniform spherical shape of powder particles is observed. The laser cladding surface modification method would overcome the limitation and enhance the wear and frictional resistance. In 316L cladding layers, small traces of d-ferrite are observed. Different phases, such as ferrite, martensite, austenite, and chrome carbides, such as $Cr_{23}C_6$ and Cr_7C_3 precipitates, are formed in the cladding region. These phases improve the hardness and wear resistance and reduce the coefficient of friction.

Fig. (8). SEM image of steel 316L powder.

Table 8. Composition of steel 316L powder

Elements	Fe	Cr	Ni	Si	Mn	Mo	C
Wt (%)	Bal.	17.1	12.8	1.0	2.0	2.2	0.1

Shahir Mohd Yusuf *et al.* [14] studied the wear behavior of 316 L steel made by additively manufacturing technique *via* laser powder bed fusion. The ultrafine grains were obtained from high-pressure processing techniques, which were used in the powder bed fusion method. The pin on disk wear test was used to perform the tribological test under the dry sliding condition, and the results demonstrated that the wear resistance of treated 316 L steel had been improved due to the fine microstructure and carbide formation. It has a consistently lower mass loss, wear rate and coefficient of friction than the as-received material. The enhancement in wear resistance was attributed to the significant improvement in hardness due to the nano-size grain refinement by the laser treatment process. Errico V *et al.* [15] studied the behavior of AISI 316L steel powder on AISI 304 steel plates. The laser power, laser spot diameter, powder feed rate and translation speed affected the track width and penetration depth. The porosity was affected by the gas flow rate and laser spot diameter. The result confirmed that the AISI 316L steel cladding surface has higher strength than the substrate surface.

Steel 420

AISI Steel 420 material is recommended to be used as injection mold due to the high corrosion resistance to vapor, carbonates, water, and other salts. These parts are usually manufactured by the electrical discharge machining process. However, industrial applications require the injection mold with the complex cooling channel, which is limited by the lack of machining processes and cost-effective approaches. Hence, the laser surface modification method is preferred to fabricate the high-quality steel 420 injection mold. The chemical composition of Steel 420 powder is shown in Table 9. Gas atomized AISI steel 420 powder is shown in Fig. (9), which displays a spherical shape that is used as a raw material for the laser cladding process. Martensite transformation occurs in the laser cladding stainless steel 420 material due to the rapid cooling process. Laser parameters are optimized to retain the martensite structure in the cladding region. The cladding region possesses both cellular and columnar dendrite structures that grow in the build direction. Different morphology of austenite can be observed by varying the heat input. $M_{23}C_6$ carbide precipitate at the grain boundary in the form of fine particles, which gives more hardness to the clad region. The shape and size of the carbide particle play an important role in determining the properties, such as toughness, wear resistance, hardness and corrosion resistance. Spheroidal shape carbide provides more hardness than other shapes of carbide.

Fig. (9). SEM image of steel 420 powder.

Table 9. Composition of steel 420 powder

Elements	Fe	Cr	Si	Mn	C	P	S
Wt (%)	Bal.	13.0	1.0	1.0	0.15	0.04	0.03

Xiao Zhao *et al.* [16] fabricated the AISI steel 420 using the selective laser melting process for the application of plastic injection mold. The characteristic of the laser melted steel 420 powder, such as phase composition microstructure and hardness, were analysed. The results showed that the hardness of the SLM material had a hardness of 50.7 HRC, which was higher than that of the casted steel 420 material. Moreover, the SLM material met the exact requirement to make the plastic injection mold. Navid Nazemi *et al.* [17] coated the steel 420 powder on the steel 420 substrate through the laser cladding process. The hardness of the cladding region was studied by varying the overlapping condition of the laser parameters. The results revealed that the hardness of the cladding region was found higher than the substrate due to the reduction in the grain size and formation of the hard face carbide.

Titanium Alloy Powder

Titanium and its alloys have been used in many industries like biomedical, aerospace and chemical industries due to high specific strength, good biocompatibility and excellent corrosion resistance. Ti6Al4V (Ti-64) is one of the titanium alloys, which has duplex structures, are widely used in different industries. The chemical composition of Ti-64 powder is shown in Table **10**. Gas atomized Ti-64 powder is shown in Fig. (**10**). The mechanical properties, such as hardness and wear resistance are dependent on the microstructure of the fabricated Ti-64 alloy. The Ti-64 powder contains Ti, Al and V elements, which have different phases depending on the heat input and cooling rate during the fabrication process. Refining the microstructure by proper selection of the process is one of the effective methods to obtain the required properties of the component. The acicular α-structure of Ti-64 can improve its microhardness using laser surface modification techniques. The hardness and wear resistance of Ti-64 can be enhanced by forming the fine phases. A basket weave substructure, a fine lamellar, lath-like lamellar phase, is the different form of microstructure that can be formed in the cladding region that defines the property of the Ti-64 alloy. The width of the different phases ranging from nanometers to several micrometers depends on the process parameter. It is becoming more and more popular for the fabrication or repair of titanium alloy components. The laser surface modification techniques have the characteristics such as steep temperature gradient, high cooling rate due to rapid cooling of the molten pool and non-equilibrium solidification. These features lead to the formation of special microstructures and the corresponding increase in mechanical properties compared to other manufacturing techniques.

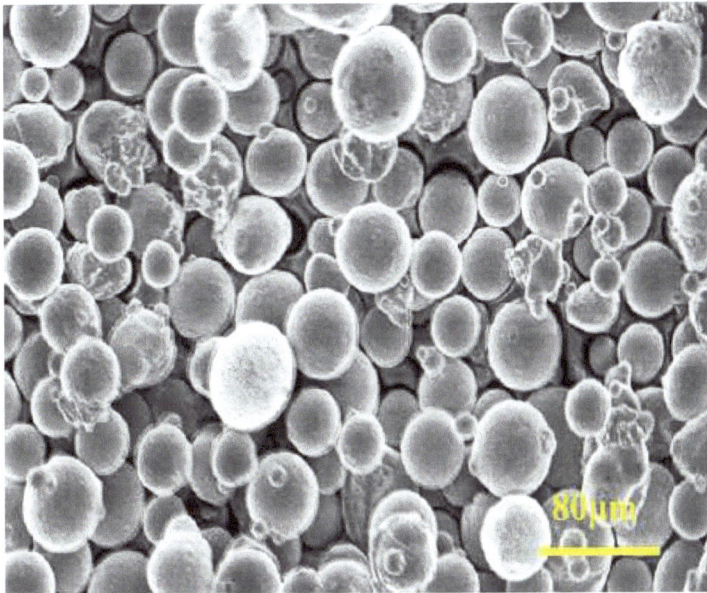

Fig. (10). SEM image of Ti-64 powder.

Table 10. Composition of Ti-64 powder

Element	Ti	Al	V	O	C	Ni	Fe
Wt (%)	Bal.	5.8	3.6	0.2	0.08	0.05	0.3

Yanqin Li *et al.* [18] used the Ti-64 hardfacing alloy powder for the laser-based additive manufacturing process. The Ti-64 microstructure was refined through the proper selection of process parameters. The microstructure, hardness, wear performance, and molten pool thermal behavior of the Ti-64 clad region were studied. The results revealed that the formation of grain sizes depended on the molten pool cooling rate. Fine grains and acicular phases were obtained at the energy density of 75 J/mm^2 due to the high solidification rate of the molten pool. Besides, the as-fabricated Ti-64 sample with acicular phase showed a higher microhardness of 7.43 GPa and elastic modulus of 133.6 GPa with a low CoF of 0.48. It is concluded that the microstructure and mechanical performance can be improved by cladding the Ti-64 alloy. Gorunov A I [19] fabricated the Ti-64 component using Ti-64 powder through an ultrasonic aided direct material deposition. The formation of grain size in the structure-built parts was studied to assess the influence of hardness and wear resistance. Tribo-wear test at high loads condition was performed, and the results showed that the ultrasonic aided laser-treated sample has lower CoF than that of the as-received material due to the formation of irregular needle structure and fine grains at the laser-treated region.

AlSiMg Powder

In AlSiMg powder, Mg is used to form the Mg_2Si precipitates for enhancing the mechanical strength of the AlSiMg based components. The chemical composition of AlSiMg powder is shown in Table **11**. The SEM image of the Gas atomized AlSiMg powder is shown in Fig. (**11**). Si is used in the form of eutectic composition that has a mixture of Al and Si for reducing the co-existence of the solid and liquid phases during the solidification process. It minimizes the defects and voids owing to volume shrinkage. Moreover, Si addition to Al alloy would increase the mechanical strength of the composition. Also, Si exhibits a high microhardness value, enhances the mechanical properties and lowers the thermal expansion coefficient. As a result, Al-Si alloys are used in cylinder heads, valve lifters and pistons in automotive applications.

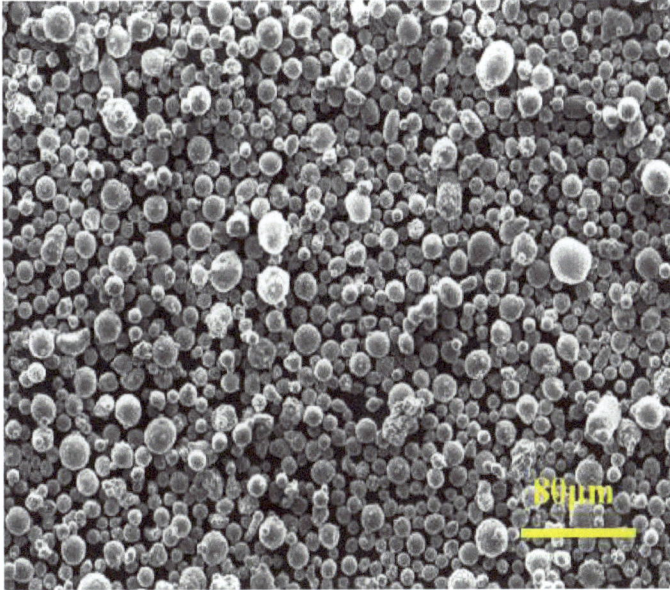

Fig. (11). SEM image of AlSiMg powder.

Table 11. Composition of AlSiMg powder

Elements	Al	Si	Mg	Fe	Mn	Zn	Cu
Wt (%)	Bal.	11.0	0.5	0.5	0.4	0.10	0.05

Kuan Jen Chen *et al.* [20] investigated the microstructure and tribological properties of Al–10Si–0.5Mg alloy manufactured through selective laser melting. Ultrafine crystals formed in the fabricated alloy sample due to the high cooling rate. It resulted in high strength, low elongation and high hardness values. Crack initiation in the fabricated parts was restricted by the uniform distribution of Si particles. The fabricated AlSiMg specimen exhibits high wear resistance and is effectively applied in the high-temperature application. Daniel Knoop *et al.* [21] studied the microstructure and mechanical properties of $AlSi_{3.5}Mg_{2.5}$ alloy, which was fabricated through laser powder bed fusion technique. Their results revealed that the tensile strength of 484 MPa and elongation at the break of 10.5% were achieved due to the formation of a supersaturated α-aluminum matrix with the fine network eutectic Si. These high-strength Al-based alloys can be utilized in automobile industries for lightweight applications.

Scalmalloy Powder

The mixture of aluminum, magnesium, scandium and zirconium elements (Al-Mg-Sc-Zr) is often referred to as Scalmalloy, which is developed to use in aerospace structural applications. Moreover, it is used in automotive for heat exchangers, robotics for hydraulic manifolds, aerospace for satellite panels and motorsports for brackets. The chemical composition of Scalmalloy powder is tabulated in Table **12**. Gas atomized Scalmalloy Powder (SEM image) is shown in Fig. (**12**). It is an Al-based alloy powder that has high strength, good welding ability, high corrosion resistance and improved elongation. Component produced by using Scalmalloy powder *via* laser source is reported to have limited residual stress and negligible internal defects. It can replace the Ti element in aerospace applications due to its high strength to weight ratio. The rapid cooling rates of the molten pool during laser treatment create the proper thermodynamic condition for a crack-free fine-grain Al structure with Al_3Sc fine precipitates. The Scalmalloy had 1.7 times the tensile strength of $AlSi_{10}Mg$ due to Al_3Sc precipitation formation. The Al-Sc mixture has a high strength to weight ratio, and Sc improves weldability in Al alloys that reduce the formation of hot cracking. Moreover, the Al-Sc mixture has high solid solution hardening and improves corrosion resistance. The increase in Sc elements would increase the tensile strength, hardness and yield strength which has low ductility.

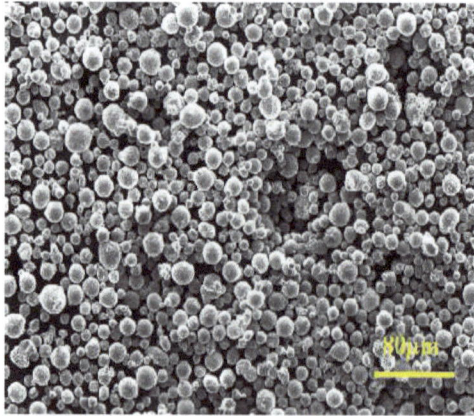

Fig. (**12**). SEM image of Scalmalloy powder.

Table 12. Composition of Scalmalloy powder

Elements	Al	Mg	Sc	Zr	Mn	Si	Fe
Wt (%)	Bal.	4.9	0.8	0.2	0.8	0.4	0.4

Zhenglong Lei *et al.* [22] studied the influence of energy density on microstructure and microhardness of 7075 aluminum alloy modified with Sc and Zr elements. The samples were manufactured by the selective laser melting process. The density of the samples increased first with the increase of energy density and then decreased. The roughness of the specimen decreased with an increase in energy density due to the increase in heat input. The microhardness of the laser-treated component was higher than that of the component prepared through conventional methods due to the nucleation of $Al_3(Sc, Zr)$ composition and grain refinement. Spierings A B *et al*. [23] have performed an additive manufacturing process *via* selective laser melting techniques to fabricate the Scalmalloy alloy sample. This sample has an advantage over the traditional casted alloy due to variation in the microstructure formation. It was found that the grain size decreased from 1.1 μm to 600 nm with the increase of laser energy density. As a result, high strength was exhibited in the build-up direction.

CuCrZr Powder

Copper alloys have significant advantages due to their superior electrical and thermal conductivity. It is preferred in applications, such as heat exchanging material, electromobility, power generation and power distribution units. The chemical composition of CuCrZr powder is shown in Table **13**. The SEM image of the gas atomized CuCrZr powder is shown in Fig. (**13**). The Cr precipitate is formed due to its poor solubility during casting in the Cu matrix. To overcome this drawback in the convention alloying process, the CuCrZr powder is transformed into solid bulk material by depositing laser energy directly through the laser powder bed fusion method. The macroscopic structure can be prepared by depositing multilayer at the desired orientation. CuCrZr is a hardenable Cu-based alloy with a high electrical conductivity that may be used for welding nozzles, current-carrying plugs and electrodes. The Cr precipitates at the grain boundaries consist of a small amount of ZrO_2 compound. The Cr precipitate formation can be restricted by automizing the CuCrZr powder. The Cr_2Zr, Cr, and Cu_4Zr precipitates are formed during the rapid solidification process that leads to an increase in both hardness and electrical conductivity.

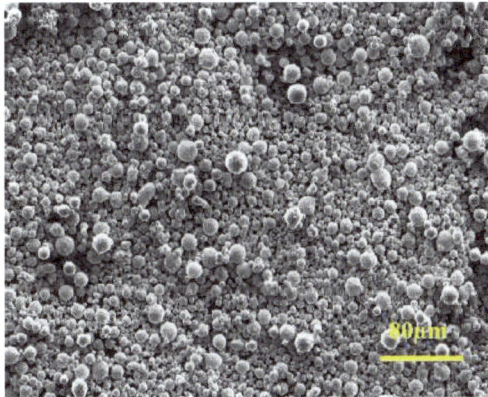

Fig. (**13**). SEM image of CuCrZr powder.

Table 13. Composition of CuCrZr powder

Elements	Ti	Al	V	O	C	Ni	Fe
Wt (%)	Bal.	5.8	3.6	0.2	0.08	0.05	0.3

Katrin Jahns *et al.* [24] focused on preparing the Cu-based alloy, which included Cu, Cr and Zr elemental powder. The CuCrZr alloy was fabricated *via* the selective laser melting process and its hardness was studied. Their result proved that the alloy prepared through the laser source has higher hardness than that of the conventional method, such as casting due to the formation of Cr-based precipitate, and formed fine microstructure. Zhibo Ma *et al.* [25] fabricated the CuCrZr alloy through the laser melting process for utilizing it in thermal and electrical conductivity. The result revealed that the sample prepared through the laser melting process has higher strength due to the formation of single-phase Cu microstructure and the strong texture was formed in the 110 plane (bcc phase), which was parallel to the built-up material direction.

CONCLUDING REMARKS

Selection of the hardfacing alloying powder for an appropriate application requires knowledge of the tribological behavior of the hardfacing alloyed component. The detail of phases which enhance the wear resistance property would help select the powder directly. This chapter described an overall view of the hardness and tribological behavior of the different hardfacing alloying powders. Different phases are formed during the hardfacing alloying/cladding process on various substrates. Despite the similar forms of wear that occur in these hardfacing components, there are some significant fundamental differences between these hardfacing components, which require assessing the phase formation to control wear. Different forms of wear are particularly controlled by the careful selection of the hardfacing alloy powder. However, hardfacing alloying powder is used to resist the adhesion wear, which may fail under abrasive or erosive wear, so that hardfacing alloying powder optimization for a particular application is essential.

CONSENT FOR PUBLICATION

Not applicable.

CONFLICT OF INTEREST

The author declares no conflict of interest, financial or otherwise.

ACKNOWLEDGEMENTS

First of all, the authors express their sincere gratitude to the editors of this book, and they extend their gratitude to their family for the continuous support provided to complete this chapter.

REFERENCES

[1] Corchia M, Delogu P, Nenci F, Belmondo A, Corcoruto S, Stabielli W. Microstructural aspects of wear-resistant stellite and colmonoy coatings by laser processing. Wear 1987; 119(2): 137-52.
[http://dx.doi.org/10.1016/0043-1648(87)90105-0]

[2] Gnanasekaran S, Padmanaban G, Balasubramanian V, Kumar H, Albert SK. Laser Hardfacing of Colmonoy-5 (Ni-Cr-Si-BC) Powder onto 316LN Austenitic Stainless Steel: Effect of Powder Feed Rate on Microstructure, Mechanical Properties and Tribological Behavior. Lasers Eng 2019; 1: 42.

[3] Zhang H, Shi Y, Kutsuna M, Xu GJ. Laser cladding of Colmonoy 6 powder on AISI316L austenitic stainless steel. Nucl Eng Des 2010; 240(10): 2691-6.
[http://dx.doi.org/10.1016/j.nucengdes.2010.05.040]

[4] Jeyaprakash N, Yang CH, Tseng SP. Wear Tribo-performances of laser cladding Colmonoy-6 and Stellite-6 Micron layers on stainless steel 304 using Yb: YAG disk laser. Met Mater Int 2019; Nov 13: 1-4.

[5] Paul CP, Jain A, Ganesh P, Negi J, Nath AK. Laser rapid manufacturing of Colmonoy-6 components. Opt Lasers Eng 2006; 44(10): 1096-109.
[http://dx.doi.org/10.1016/j.optlaseng.2005.08.005]

[6] Feng K, Chen Y, Deng P, *et al.* Improved high-temperature hardness and wear resistance of Inconel 625 coatings fabricated by laser cladding. J Mater Process Technol 2017; 243: 82-91.
[http://dx.doi.org/10.1016/j.jmatprotec.2016.12.001]

[7] Anil PM, Naiju CD. Sliding Wear Reliability Studies of Inconel 625 Components Manufactured by Direct Metal Deposition (DMD). Procedia Manuf 2019; 30: 581-7.
[http://dx.doi.org/10.1016/j.promfg.2019.02.082]

[8] Samuel C, Arivarasu M, Prabhu TR. High temperature dry sliding wear behaviour of laser powder bed fused Inconel 718. Addit Manuf 2020; Aug 1: 34.

[9] Zhao Z, Qu H, Bai P, Li J, Wu L, Huo P. Friction and wear behaviour of Inconel 718 alloy fabricated by selective laser melting after heat treatments. Philos Mag Lett 2018; 98(12): 547-55.
[http://dx.doi.org/10.1080/09500839.2019.1597991]

[10] Li B, Jin Y, Yao J, Li Z, Zhang Q, Zhang X. Influence of laser irradiation on deposition characteristics of cold sprayed Stellite-6 coatings. Opt Laser Technol 2018; 100: 27-39.
[http://dx.doi.org/10.1016/j.optlastec.2017.09.034]

[11] Telasang G, Majumdar JD, Wasekar N, Padmanabham G, Manna I. Microstructure and mechanical properties of laser clad and post-cladding tempered AISI H13 tool steel. Metall Mater Trans, A Phys Metall Mater Sci 2015; 46(5): 2309-21.
[http://dx.doi.org/10.1007/s11661-015-2757-z]

[12] Telasang G, Majumdar JD, Padmanabham G, Tak M, Manna I. Effect of laser parameters on microstructure and hardness of laser clad and tempered AISI H13 tool steel. Surf Coat Tech 2014; 258: 1108-18.
[http://dx.doi.org/10.1016/j.surfcoat.2014.07.023]

[13] Fouquet F, Sallamand P, Millet JP, Frenk A, Wagniere JD. Austenitic stainless steels layers deposited by laser cladding on a mild steel: realization and characterization. J Phys IV 1994; 4(C4): C4-C89.
[http://dx.doi.org/10.1051/jp4:1994418]

[14]   Yusuf SM, Lim D, Chen Y, Yang S, Gao N. Tribological behaviour of 316L stainless steel additively manufactured by laser powder bed fusion and processed *via* high-pressure torsion. J Mater Process Technol 2021; 290: 116985.
[http://dx.doi.org/10.1016/j.jmatprotec.2020.116985]

[15]   Errico V, Campanelli SL, Angelastro A, Mazzarisi M, Casalino G. On the feasibility of AISI 304 stainless steel laser welding with metal powder. J Manuf Process 2020; 56: 96-105.
[http://dx.doi.org/10.1016/j.jmapro.2020.04.065]

[16]   Zhao X, Wei Q, Song B, *et al.* Fabrication and characterization of AISI 420 stainless steel using selective laser melting. Mater Manuf Process 2015; 30(11): 1283-9.
[http://dx.doi.org/10.1080/10426914.2015.1026351]

[17]   Nazemi N, Alam MK, Urbanic RJ, Saqib S, Edrisy A. A hardness study on laser cladded surfaces for a selected bead overlap conditions. SAE Technical Paper 2017.
[http://dx.doi.org/10.4271/2017-01-0285]

[18]   Li Y, Song L, Xie P, Cheng M, Xiao H. Enhancing hardness and wear performance of laser additive manufactured Ti6Al4V alloy through achieving ultrafine microstructure. Materials (Basel) 2020; 13(5): 1210.
[http://dx.doi.org/10.3390/ma13051210] [PMID: 32182664]

[19]   Gorunov AI. Additive manufacturing of Ti6Al4V parts using ultrasonic assisted direct energy deposition. J Manuf Process 2020; 59: 545-56.
[http://dx.doi.org/10.1016/j.jmapro.2020.10.024]

[20]   Chen KJ, Hung FY, Lui TS, Tsai CL. Improving the applicability of wear-resistant Al–10Si–0.5 Mg alloy obtained through selective laser melting with T6 treatment in high-temperature, and high-wear environments. J Mater Res Technol 2020; 9(4): 9242-52.
[http://dx.doi.org/10.1016/j.jmrt.2020.06.078]

[21]   Knoop D, Lutz A, Mais B, von Hehl A. A tailored AlSiMg alloy for laser powder bed fusion. Metals (Basel) 2020; 10(4): 514.
[http://dx.doi.org/10.3390/met10040514]

[22]   Lei Z, Bi J, Chen Y, Chen X, Qin X, Tian Z. Effect of energy density on formability, microstructure and micro-hardness of selective laser melted Sc-and Zr-modified 7075 aluminum alloy. Powder Technol 2019; 356: 594-606.
[http://dx.doi.org/10.1016/j.powtec.2019.08.082]

[23]   Spierings AB, Dawson K, Uggowitzer PJ, Wegener K. Influence of SLM scan-speed on microstructure, precipitation of Al3Sc particles and mechanical properties in Sc-and Zr-modified Al-Mg alloys. Mater Des 2018; 140: 134-43.
[http://dx.doi.org/10.1016/j.matdes.2017.11.053]

[24]   Jahns K, Bappert R, Böhlke P, Krupp U. Additive manufacturing of CuCr1Zr by development of a gas atomization and laser powder bed fusion routine. Int J Adv Manuf Technol 2020; Mar 16: 1-1.
[http://dx.doi.org/10.1007/s00170-020-04941-7]

[25]   Ma Z, Zhang K, Ren Z, Zhang DZ, Tao G, Xu H. Selective laser melting of Cu–Cr–Zr copper alloy: Parameter optimization, microstructure and mechanical properties. J Alloys Compd 2020; 828: 154350.
[http://dx.doi.org/10.1016/j.jallcom.2020.154350]

CHAPTER 4

Lasers and their Industrial Applications

Venkata Charan Kantumuchu[1, *]

[1] *American Society for Quality, Oklahoma City, Oklahoma, the United States/Mauser Packaging Solutions, Memphis, Tennessee, USA*

Abstract: LASER expanded as Light Amplification by Stimulated Emission of Radiation is one of the greatest inventions of mankind. The history of lasers dates back to the early 1960s. There are various types of lasers, such as gas lasers, chemical lasers, dye lasers, semiconductor lasers, solid-state lasers, CO_2 lasers, Nd-YAG, and fiber lasers, *etc.*. The applications of lasers are many and are only growing with time as new ways of laser applications are uncovered. Some critical applications of lasers are mechanical, medical, aviation, automotive, fabrication, sheet metal, electronics, electrical, packaging, and tribology treatment. Within these applications, lasers are used for a wide variety of reasons. The global laser processing market size was valued at $12.5 billion in 2017. The industrial lasers market is expected to witness a CAGR growth of 5.3% until 2023. The chapter also researches how the quality of laser treatments affects the market.

Keywords: Aerospace, Applications, Automotive, Barcoding, Biomaterials, Data storage, Fashion technology, Healthcare, Lasers, Laser cutting, Laser lights, Market size, Matrix laser, Microprocessors, Printing, Quality, Satellite communication, Scanning, Semiconductor, Skin treatments, Types of Lasers.

INTRODUCTION

LASER is an abbreviation of "Light amplification by stimulated emission of radiation," and it is one of the greatest inventions of mankind. As the definition suggests, it is the stimulated emission of light that generates the laser. Without getting into the intricate technical details, the process may be explained in the following three steps: light absorption, spontaneous emission, and stimulated emission. Apart from spontaneous emission, which was already known for quite some time, from observation of fluorescence, Albert Einstein postulated the process of stimulated emission in 1917 [1]. However, the use of lasers was started in the 1960s. The first laser was operated on May 16, 1960, by Theodore Maiman at Hughes Research Laboratory [2]. The laser solves various problems; lasers

---

* **Correspondence author Venkata Charan Kantumuchu:** American Society for Quality, Oklahoma City, Oklahoma, USA; E-mail: charan.kantumuchu@mauserpackaging.com, Tel: +13179975641, Fax: 05043160402;

Jeyaprakash Natarajan and Che-Hua Yang (Eds.)

have various applications in technology, aerospace, automotive, medical, health-care, manufacturing, entertainment, and logistics. In simple terms, a laser can be considered as a special flashlight, with some clear differences. There are three distinct differences between a flashlight and a laser. The first difference is that the laser beams are much narrower than flashlight beams (collimation). The second is that the white light of a flashlight beam contains different colors of light colors, while the laser beam only contains one color (monochromatic). The third is that the light waves in a laser beam are aligned with each other (high coherence); however, the flashlight waves are randomly arranged [3]. These abilities, along with others, make lasers a choice of application from supermarkets to eye surgeries. This chapter will elaborate on the various applications of lasers, along with some case studies, as applicable, to understand the applications better. The chapter will also discuss how the quality of laser application compares to a conventional method. The application of lasers for tribology treatment and how lasers will improve the frictional wear due to the parts moving in relation to each other are discussed in detail in other chapters. Hence, we will cover these applications only superfacially in this chapter. It is estimated that approximately 23% of the global energy is consumed by tribology [4], which emphasizes the seriousness of the matter. This analysis shows that there is a scope to save millions, if not billions of dollars globally.

CLASSIFICATION OF LASERS

The applications of lasers vary with the types of lasers and their limitations. Hence, it is essential to discuss the types of lasers for a better understanding of their applications. There are various types of lasers. Lasers may be classified in different ways. Lasers may be classified according to the physical state of the active material, the wavelength of the emitted radiation, and even the output powers [5]. Here is a category of lasers based on the active material (Fig. **1**) [6]:

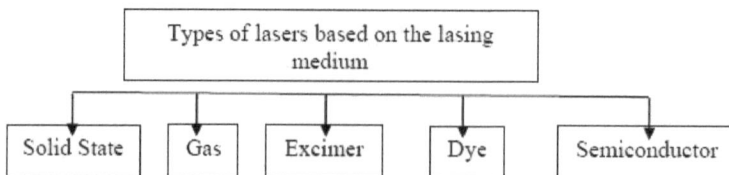

Fig. (1). Flow chart of types of lasers based on the active material.

1. Solid-state
2. Gas
3. Excimer
4. Dye
5. Semiconductor

Solid-state: These lasers have a solid matrix lasing medium (example: Nd-YAG – neodymium-yttrium aluminum garnet).

Gas lasers: These lasers have gases as lasing medium (example: helium, helium-neon, and CO_2 lasers).

Excimer lasers: These lasers use excited dimers as the lasing medium (example: chlorine and fluorine mixed with inert gases like argon and krypton).

Dye lasers: These lasers use complex organic dyes as the lasing medium (example: rhodamine 6G in liquid solution).

Semiconductor lasers: These are not solid-state lasers and function by electric current flow to a semiconductor (Example: Lead salt).

Because of the wide variety of applications of lasers, the market for lasers is worth billions of dollars. In 2017, it was estimated that the laser market was worth $12.5 B, and this market will experience a CAGR growth of 5.3% until 2023. The most critical market for lasers is communications at 4 billion euros, macro material processing at 2.3 billion, and microprocessing at 1.3 billion euros. In addition, the medical market is at 100 million euros, sensors and instruments at 800 million and marking at 645 million euros [7]. These numbers highlight the potential and prospective applications that are increasing exponentially year over year.

INDUSTRIAL APPLICATIONS

As discussed at the beginning of the chapter, in the following sections, the chapter will focus on the industrial, automotive, aerospace, medical and other applications of lasers in detail with case studies where essential. The chapter will highlight the differences between conventional methods and laser applications for achieving the desired output. There are numerous industrial applications of lasers. In this section, we will cover some of the more common applications of lasers for industrial applications.

Laser Printing

Laser printing is a revolution in manufacturing. Its applications span from manufacturing to medicine. "Laser printing" is a general term that expands to rather a large set of additive direct writing techniques [8]. LIFT, which stands for laser-induced forward transfer, is a technique that works according to the principle of operation that a controlled amount of material is transferred from a donor system to a receiving substrate employing laser irradiation [9]. In these approaches, a pulsed laser is used to induce the transfer of material from a source film onto a substrate in close proximity to or in contact with the film. The source is typically a laser-transparent substrate coated with the material of interest, referred to in the literature as the target, donor, or ribbon. Laser pulses propagate through the transparent substrate and are absorbed by the film [10].

For the process of laser printing on paper, Xerox has a five-step process. The process includes charging, exposure, development, transfer, and fusing. In the charging phase, the entire photoreceptor is negatively charged. In the exposure phase, a laser beam, according to the image of the original document, is irradiated onto the negatively charged photoreceptor. In the development phase, the toner is fixed to prevent it from coming off by applying heat and pressure to melt the toner into the paper. In the transfer phase, the image of the original document is transferred onto the paper. Just as in the development phase, fusing uses heat and pressure to permanently bond toner on paper. This process completes the laser printing [11]. The quality of the laser printers can be understood better by comparing the laser printers with the inkjet printers. When the quality of the printers is considered, several factors could be taken into consideration. However, for this chapter, let us consider the print quality, print speed, print volume, page yield, printer span, and cost per page [12].

Similarly, there are other types of laser-assisted printing, such as laser printing of ceramics and polymers, soft materials, nanoparticles, electronic materials, chemical and biological sensors, proteins and biomaterials, laser-assisted bioprinting of cells for tissue engineering, industrial digital printing, and laser printing of 3D metal structures [9]. Table **1** shows the differences between inkjet *vs.* laser printers.

Table 1. Showing some differences between inkjet *vs.* laser printers.

Printer Types Characteristics	Inkjet Printer	Laser Printer
Print quality	Brilliant color pictures	Sharp documents
Print speed	16 pages per minute (ppm)	15 to 100 pages per minute (ppm)
Print volume	250-500 pages @15 ppm Canon PIXMA TS6220	750-3000 pages @35 ppm HP LaserJet Pro M401n
Page Yield	Most ink cartridges usually 135-1000 pages	Toner cartridge from 2000-10000 pages
Printer Span	A minimum span of 3 years	A minimum span of 5 years
Cost per page	6.2 cents per page	2.8 cents per page

Lasers in 3D Printing

3D printing is certainly one of the fastest-growing technologies with applications in several areas, including industrial, medical, automotive, aviation, and electronic industries. There are various 3D printing technologies, and stereolithography is one of them. The first 3D printing technology to be invented was based on photopolymerization called stereolithography (SLA). A stereolithographic apparatus uses a computer-controlled laser beam to build a 3D object within a liquid photopolymer tank. This layer-by-layer printing involves curing a liquid photopolymer that forms the desired 3D object [13]. Selective laser sintering or SLS is another 3D printing technology that uses a laser to print 3D objects. SLS uses a laser with selectively stratified sintering solid powder and the sintering layers to generate the required shape parts [14]. As discussed before, stereolithography has many applications, but one of the most critical applications is perhaps the bio-medical engineering field. SLA enables surgeons to make models of patients and implants to practice sensitive procedures in advance [15]. This method helps the surgeons, patients, and families understand the process better and make an informed decision.

SLS, like SLA, has various applications in aviation, consumer goods, and automotive industries. For example, SLS may be used to create several parts in a flight cabin, such as video monitoring shrouds and air vent grills. In addition, millions of 3D printed mascara brushes are being used by Chanel in the beauty and fashion industry using SLS. Furthermore, Alfa Romeo leveraged SLS 3D printing to print aerodynamic wind tunnels, laminating molds, cores, and complex serial parts [16]. 3D printing using lasers (SLA and SLS) makes it possible to produce parts at lower costs demonstrating several aspects of lean manufacturing and quality. Because of the ability to produce parts just in time and incurring almost no setup costs, 3D printing can help save millions, if not billions of dollars. At the beginning of 2011, there were $537 billion in inventories in the manufacturing industry, equal to 10% of that year's revenue [17]. This adds to other costs, such as over-processing, defects, and transportation. These costs would make up a significant portion of the money held for non-value-added activities in various industries, and 3D printing can eliminate/minimize many of these wastes.

Lasers in Manufacturing

There are various manufacturing processes, such as casting, forging, rolling, extrusion, molding, welding, and also machining processes, such as turning, milling, shaping, planning, and drilling. Traditional manufacturing has been in use for several centuries. However, the traditional manufacturing processes have some drawbacks for specific applications. For example, laser forming is an excellent choice for the rapid fabrication of prototypes since conventional manufacturing involves high design times and the cost of dies, making it uneconomic. Similarly, laser welding is preferred over conventional welding because of the various advantages, such as high welding speeds, the precision of weld, and the slight heat affected area.

On the other hand, the traditional machining processes work by removing the excess material using a tool when it comes to machining. However, there are various applications where traditional machining is either inefficient or not economical. For this reason, laser machining is used widely since laser machining can produce parts with high precision where conventional machining processes cannot and yet be more economical. Lasers are now widely used in the industries for various machining operations, such as drilling, cutting, and shaping materials. "The laser machining operations are efficient and economical in many industrial applications where large production rates are desired. In addition, laser drilling is beneficial for high-aspect-ratio micro drilling applications where conventional mechanical drilling is not applicable or less efficient" [18]. Table **2** shows some significant benefits of using laser drilling [19].

Table 2. A comparison of conventional *vs.* laser drilling.

Characteristic	Conventional Drilling	Laser Drilling
Heat	The heat produced from friction between the tool and the material removed	Minimized heat effect (virtually no recast layer)
Precision and depth	Depends on the drill bit	Ten times smaller than the smallest drill bit
Speed	Depends on the drill bit and the material	Up to 10,000 locations per second
Aspect ratio	Depends on the drill bit and the material	30:1 in some materials
Drilling ability	Limited by material hardness	Based on the absorption of the target material

Within the manufacturing industry, it is estimated that approximately 40% of frictional losses are because of tribological implications. According to research conducted on the effect of laser surface texturing (LST) on hot stamping dies, it was found that "Laser texturing is a suitable process to improve the tribological performance of hot stamping dies since all the textures evaluated in this work was better than the untextured surface to prevent metal-to-metal contact" [20].

Similarly, laser texturing of both coated and uncoated metals has significant improvement regarding tribological properties. LST creates micro dimples (craters), which act as reservoirs for the lubricant and also as a trap for the debris, thereby increasing the life of the moving parts [21]. "The lifetime of the structured samples, defined as the sliding distance after which the friction coefficient showed an abrupt increase, was found to be significantly enhanced for all structured surfaces. The enhancement ranged from a 30% lifetime increase when TiN coatings were directly structured to an increase of over ten times for structured hard metal which was subsequently coated with TiCN" [22].

COMMERCIAL APPLICATIONS

Apart from manufacturing and 3D printing, there are other laser applications in the commercial market, some of which will be discussed in this section. The following applications use lasers in their basic functionality.

BAR Code Scanners and Printers

It is not uncommon to come across bar code scanners in supermarkets, malls, inventory management systems in small and large scale industries (SKUs), admission tickets of entertaining events like movies and theme parks, air travel, advertising, and various other places where quick identification is required. Bar code scanners not only save time but also make several processes seamless. Bar codes eliminate human errors, are inexpensive, do not require a particular skill to operate, and instantaneously retrieve data that are helpful in quickly making decisions. Giant companies like Walmart and Amazon have realized the importance of using bar code scanning technology. Walmart had started using bar codes at their checkouts as early as 1983. Today, they have leveraged this technology to help their customers get the best price through their "Savings catcher" program [23].

Similarly, an Amazon distribution warehouse covering one million square feet unloads 100 trucks with over 1200 tons of items at 22 docking bays, scanning each item to a specific location in the warehouse [24]. A typical manual unloading and sorting system could take about 15-20 seconds to identify and segregate an item, while the barcode scanners do that in 2-3 seconds. This is almost 85% faster. It is reasonably simple to understand how critical the bar code scanners are in the "one/two-day shipping," if not "same day shipping," logistic abilities of thousands of companies, with the speed, efficiency, and precision they offer.

CD, DVD and Optical Discs

Many of us may have used or seen a CD at some point in time. CDs were once a great way to store music, movies, videos, and more. It was not unusual to have a collection of CDs of music or movies, and many people still do. Though CDs are not as popular now, they have replaced the bulky and relatively brittle vinyl records. The compact disc technology was invented in 1979, and since its introduction to the public in 1982, it was estimated that 200 billion CDs had been sold worldwide [25]. Some readers may still recollect how using magnetic tape cassettes was not so versatile, as the listener had to either rewind or fast forward to a specific song mechanically. CDs have introduced the ability to play a specific song or video within a fraction of its time on magnetic tapes. From a quality perspective, CDs were much more robust than magnetic cassettes. They could last for several years with minimum maintenance, while the magnetic tapes could deteriorate in quality without the required maintenance. Also, the quality of music and video content on optical discs is far superior to magnetic tapes. Finally, the data stored on optical discs is many-fold compared to the data stored on magnetic tapes.

DVDs and other optical discs like Blu-ray technology had come into the picture, eventually. Today, the CD technology is replaced by either mass storage devices like a thumb drive or cloud technology, which could avoid the hassle of carrying a separate device altogether. In addition, there are various streaming services for both music and movies, like Amazon, Hulu, Spotify, Pandora, Sirius XM, Apple Music, YouTube, and more [26]. However, compact disc technology has helped millions of people worldwide for almost 40 years (and still does), who enjoy music, movies, data, and more.

These are some of the industrial applications of lasers, and there are many more. While there are many more other applications like the lasers used in live concerts, lasers in technology like OLED display manufacturing, laser pointers, lasers for accurate measurement, optical communication, and laser tag games that can be discussed, they remain outside the scope of this chapter and hence are not discussed here. Some of these applications may be covered in the automotive, aviation, and medical applications.

LASERS IN THE AUTOMOTIVE INDUSTRY

The application of lasers has spanned into several industries, along with the automotive industry. There are various ways the automotive industry puts lasers to use. This section will discuss some of the general applications of lasers in the automotive industry and then discuss in detail the laser applications at some of the OEMs. We have already considered some applications like laser welding, laser cutting, laser drilling, and other laser machining processes in the previous sections. These processes are used in the automotive industry as well. Laser welding and laser cutting are standard processes used in the automotive industry because of the various benefits. For example, laser welding has benefits, such as high welding speeds, reduced HAZ, no flow of current, and more accurate welding. Laser welding has rapidly replaced traditional welding methods, especially in the automotive industry. "Resistance welding is probably the technique that has most easily been replaced by laser welding, particularly in the automotive and domestic goods industries. Laser welding is competitive primarily when high productivity, high joint quality, and low distortion are required. In the early 1990s, it was estimated that North American automobile manufacturers could save one thousand million dollars annually by using laser processing, with savings at least $100 per car through changes to the bodywork alone. In Japan, more than one-quarter of all CO_2 lasers delivered are destined for the automotive industry. The growth of laser welding in the automotive industry has been spectacular to the extent that virtually every automobile manufactured in high volumes now contains laser-welded parts in all the product categories" [27]. This is an excellent example of how laser welding has transformed the automotive industry into more efficient and economical. Ford has turned to laser welding for the mass production of Ford Mustang to avoid cutting holes in the hydroformed tube for the traditional resistance spot welding as cutting holes on both sides would create structural weakness [28]. There are several advantages to using laser welding *vs.* traditional welding methods like MIG welding from a quality perspective. MIG can cause structural or metallurgical damage due to widespread heat transfer. The laser reduces the heat input by roughly 85% compared to MIG, which will also alleviate the residual stresses induced by the heat input. Finally, in the latest generation of high-strength steels, the strength of the steel is compromised when melted and solidified at a low cooling rate, like in the case of MIG welding and laser can help maintain the parent strength [29].

Another vital application of lasers in the automotive industry is laser cutting. "Lasers are routinely used in the manufacture of automobiles. Laser cutting of stainless steel, nickel alloys, and other metals find widespread application in the aircraft and automobile industry" [30]. Magna, one of the biggest automotive suppliers in North America, has attributed its success in winning the "Supplier of the Year - Innovation award" to the unique process of laser cutting and welding front fascia on Camaro ZL1. Magna also highlighted that the laser cutting and welding process enables a lighter-weight design and savings on investment and floor space for lower volumes [31].

One of the biggest automobile companies, Honda, has developed a new laser cutting technology that would be ten times faster than the previous laser methods. "The new method has several advantages over stamping, starting with flexibility. Stamping relies on heavy dies that are expensive and time-consuming to change when parts for a new vehicle are needed, or a design change has been introduced" [32]. Japanese carmakers are known for their versatility to maintain inventories through JIT and improve cycle times through SMED methods [32]. Similarly, other laser manufacturing methods have their advantages over traditional manufacturing methods. Again, laser cutting is a much faster, cleaner, and more economical process from a quality perspective. Both pulsed and continuous lasers have been used for laser cutting, which offers significant advantages, such as cutting complex geometries, faster processing speeds, cutting of a wide range of materials, clean cuts, and other uses [18].

Another application of lasers discussed in detail in this section is laser headlights. According to Osram, one of the biggest producers of automotive lighting, "Laser light is an absolute innovation in automotive lighting and the next big step forward since the introduction of halogen, xenon, and LED headlight technologies." Laser diodes are approximately ten times smaller than LEDs, yet they have high luminance, which is far more than the light sources today. Laser lights offer up to a 20% greater headlight range relative to the current standard of the LED high beam systems" [33]. BMW has used laser headlights in their iconic BMW i8 series. According to an engineer at BMW, laser headlights use 30% less energy than LED lights while providing the same photometric performance for the same size headlight. The laser light is also expected to increase the visibility, covering up to 660 yards, while being several times smaller than the xenon and halogen headlights [34].

On the other hand, Audi has developed a new laser technology named matrix laser technology. The new technology operates with a rapidly moving micro-mirror, which redirects the laser beam. As a result, the light is distributed to a larger projection area at low vehicle speeds, and the road is illuminated with an extensive range. At high speeds, the aperture angle is smaller, and the intensity and range of the light are increased significantly. This increases visibility and safety and a more acceptable dynamic resolution and a higher degree of utilization [35].

There are various other laser applications in the automotive industry, such as laser drilling, marking, measuring, engraving, and even laser cleaning. There has been news that one of the biggest EV makers, Tesla, has planned to use lasers in place of wipers to clean the windshield. According to the news article, Tesla has developed an innovative method to use lasers to clean the windshield to ensure the autopilot cameras can function seamlessly all the time [36].

Finally, it is critical to consider how laser surface texturing (LST) is helping the automotive industry from a tribological improvement standpoint. One of the critical parts of an IC engine is piston rings. Piston rings may affect the vehicle's performance through fuel consumption, engine life, reliability, and exhaust gas composition. In an experiment conducted on the partially laser surface textured rings in an IC engine, tested on a dynamometer using a Ford Transit naturally aspirated 2500 cm^3 engine at nearly half the load and variety of speeds, the LST piston rings exhibited a 4% improved fuel efficiency (lower consumption), when compared to conventional barrel-shaped rings [37].

To understand the severity of the piston and piston ring system in generating friction, the claim discussed here is to be considered. Interestingly, about 40% of the total energy developed by an average automotive engine is consumed by frictional forces. Moreover, of this 40% frictional force, 50% to 60% of the frictional force is developed alone by the piston and the piston ring system. Therefore, a separate experiment was conducted to understand the effect of laser surface texturing *vs.* non-surface textured reciprocating automotive parts. The results showed that an actual friction reduction of 28% was observed with textured surface components compared to non-surface textured components [38].

LASERS IN THE AVIATION, SPACE, AND DEFENSE INDUSTRY

As discussed in the previous section, aviation, like the automotive sector, uses lasers for cutting, welding, machining, and other manufacturing operations. Apart from these applications, one of the practical uses of lasers in the aviation industry is their metrology application. The aviation industry needs explicitly highly reliable and precise measuring systems because of the safety regulations. An example is Airbus, which has used laser tracker measurement for automated positioning in its new fuselage structure assembly line for A320 in Hamburg, Germany. The new facility has several robots and a new logistics concept, along with the new laser measuring system, to improve the efficiency of the assembly process. According to the news article, "These will further support Airbus' drive to improve both quality and efficiency while bringing an enhanced level of digitization to its industrial production system" [39].

Laser welding, which has been discussed in the previous sections, has a significant role in aerospace. Laser combined with arc welding is a new technology called hybrid laser-arc welding. It has various benefits like deeper penetration and higher welding speeds, which neither laser welding nor arc welding can provide on their own. "In recent years, hybrid laser-arc welding has been increasingly used to join aeronautical materials; the benefit mainly originates from the ability of this process to adjust filler metal additions, heat input, and post heat treatment processes" [40]. Similarly, laser welding is also replacing TIG welding in some areas of aerospace manufacturing. "In aerospace applications, fiber laser welding has the potential to significantly add to this growth rate because of its many advantages over Tungsten Inert Gas (TIG) and Electron Beam (EB) welding processes for welding exotic aerospace alloys" [41]. It is known for aerospace applications to desire minimum residual stresses for various applications. The microscopic cracks can develop over time, and laser shock peening (LSP) is a new technology to alleviate the residual stresses. This treatment of laser peening helps extend the life of the parts, and in the case of Rolls Royce, this treatment has been used on the fan blades. "Following our comprehensive evaluation of the safety and benefits of LSP-treated blades, Rolls-Royce has adopted LSP to treat its Trent 800 engine. Since then, the company has applied LSP to Trent 500, Trent 800, Trent 1000, and XWB engines. So far, Rolls-Royce has sold more than 1,200 of its latest XWB engines – worth over £60 billion in total" [42].

Another essential and perhaps interesting application of laser in the aerospace industry is in communication. One known problem with space communication is the use of RF signals that slow down communication. "Satellites can transmit data directly to ground stations, but to do so, they require a direct line-of-sight between the satellite and the receiving ground station, which is only available for a few minutes per day" [43]. To address this problem, more efficient and reliable communication is required through the use of lasers. Satellites have a communication challenge as well that laser technology can help overcome. "Historically, satellite communications have been conducted using RF (Radio Frequency) beams. For a GEO satellite, RF is perfect since it can be groomed to cover various regions. However, RF is too wide, too dispersed for a satellite and would create interference with other satellites" [44]. Lasers can help communicate faster and reliably since lasers work in a tight, narrow beam directed precisely to a target. Speed plays a vital role in defense applications since communication delays during a war can sometimes change the situation's outcome. "Laser communications, also called optical communications, is not a new capability. The photon- or light-based technology relies on lasers to transmit data through space by a satellite. Experts venture that optical communications will provide unprecedented communication speeds, security, reliability and low latency" [45].

Department of defense has experienced similar communication challenges during the Persian Gulf War in 2003. "During the second Persian Gulf War in 2003 (Gulf War II), DOD officials relied heavily on commercial satellites to augment overwhelmed military assets. Still, officials worry those may not always be available — especially if there is a need to cover a larger geographic area than the Gulf War or an involvement of two widely separate locations" [46]. The answer is again a laser technology – lasercom. According to a claim made by Ball Aerospace, the company is building a communication technology that is several times faster than an RF signal. "The first global, all commercial optical communications satellite system is on its way, and we are using proven hardware to make it happen. Therefore, if you are looking for high bandwidth data and a next-generation service, you have just found it. We are building a satellite system to deliver space-based laser communications at speeds nearly 100 times faster than conventional radio links" [47]. The other application of lasers for defense also includes laser weapons. For example, Boeing, among other companies like Lockheed Martin and Northrop Grumman, has developed a compact laser weapons system that directs energy on its targets, focusing on one spot and damaging the target [48]. Several other uses of lasers in aerospace, space, and defense sectors are not discussed in detail here. So, to conclude, lasers and laser fabrication methods are game-changers that would improve safety, structural monitoring, cost of manufacturing and maintenance, and provide ease of fabrication and reduce operating costs [27].

Laser surface texturing is not only used in automotive and general manufacturing environments but also in aerospace. Specifically for aerospace applications, it is essential to reduce friction and the wear caused by it. Worn-out parts could translate to costly maintenance and repair activities, causing downtime and increased costs. Laser surface texturing is an excellent solution to this problem; the reservoirs formed by LST help retain the lubricant, reducing friction and wear. According to an experiment conducted, "Achieving high durability and long life requires solid lubricant reservoirs near the tribological surface. Reservoirs were fabricated using precision laser processing to generate arrays of micrometer-sized holes with well-controlled size, location, and density on the surface of hard ceramic coatings applied to steel substrates. The result indicates that there is an optimum reservoir surface density, where the life of the solid lubricant is improved by order of magnitude" [49].

According to key figures from the transportation industry, energy used to overcome friction in the road transport industry is 32%. In comparison, that of rail and sea is 20%, and air transport is estimated to be at 10% [50]. So, it is critical to mitigate the amount of energy consumed in these industries. In a different experiment conducted on a titanium alloy Ti6Al4V, which is widely used in the manufacturing of aerospace parts, the result showed a significant reduction of wear and friction, almost 99% and 67%, respectively, when textured with laser speeds of 400 mm/s and 800 mm/s [51].

LASERS IN THE MEDICAL INDUSTRY

Lasers have numerous applications in the medical sector. We may not be able to cover all the applications of lasers here in this section. However, we will discuss some critical applications, such as laser treatments of hair (both removal and growth), skin, eyes, dentistry, and medical surgeries. Both hair removal and hair growth treatments can be done with the help of lasers. The process can be simply explained as the damage of the tube-shaped sacs that produce hair through the energy emitted by a laser.

"Laser hair removal was first approved by the FDA in 1995 and has become an increasingly popular procedure over the past several years. The use of laser or light sources to remove unwanted hair is based on selective photo thermolysis theory. This concept allows the user to transfer light energy into heat and destroy the hair follicle with minimal to no effect on the surrounding tissue" [52]. According to the American Society of Aesthetic Plastic Surgery Statistics for 2019, laser hair removal was the third most performed nonsurgical procedure in 2019 with 180,332 cases. This was increased from 118,592 cases in 2018, a 52% increase in 2019 [53]. This is the growing popularity of laser hair removal treatments across the US. The treatment costs have varied from $100 to $800 per session in the U.S., depending on the city and the body part the treatment is done on [54].

The second hair procedure associated with laser is hair growth. Unlike hair removal, where the tube-shaped hair growing sacs are damaged with lasers, here, low-level laser therapy is used to invigorate circulation and stimulation that encourages hair follicles to grow hair. The treatment claims to be effective with a study done on several males and females with a 39 percent hair growth over 16 weeks [55].

There are several products in the market today that assist in the laser treatment of hair regrowth. These products vary in price depending on the number of lasers, effectivity, and the period it may take to be effective. The products work on the theory of the low-level lasers stimulating hair growth, but they may also have some proprietary technology to enhance the hair growth process. For example, the Hairmax laser comb is a product that has a patented technology of hair parting teeth mechanism that would make the laser treatment much more efficient. "The device uses a technique of parting the user's hair by combs that are attached to the device. This improves the delivery of distributed laser light to the scalp. In addition, the combs are designed so that each of the teeth on the combs aligns with a laser beam. By aligning the teeth with the laser beams, the hair can be parted and the laser energy delivered to the scalp of the user without obstruction by the individual hairs on the scalp" [56].

Similarly, a different product uses low-level light laser therapy to stimulate the scalp and then the blood flow to the scalp. Kiierr is a baseball cap that has inbuilt laser diodes that emit laser light to promote hair growth. Kiierr claims that "The low-level lights send photons into the human tissue, which are absorbed by weaker cells to help them restore their strength. Once this happens, the cells can return to their normal functionality, once again re-growing hair. To help the new growth and maintain the growth of the surrounding, healthy hair, the laser treatments increase blood flow to the scalp" [57].

Laser treatments are also available for skin conditions like skin tightening, skin resurfacing, scars removal, tattoo removal, facelift, and other conditions. Lasers are also used to treat more severe skin conditions like vascular lesions. Vascular lesions are abnormalities of skin and the underlying tissues. "Lasers for the treatment of vascular lesions are continually improving using the principles of selective photo-thermolysis. Important variables to consider in the treatment of vascular lesions are size, depth, and composition. Effective treatment of vascular lesions requires tailoring the light wavelength, fluence, and pulse duration to the specifics of the targeted lesion" [58].

Tattoo removal is another standard procedure that involves treating skin with lasers. It is estimated that nearly 30% of the American population has at least one tattoo. While this could be true, it is possible that a decent number of this population may have a change of mind for personal or professional reasons and may choose not to keep the tattoo anymore. Lasers help do this with a unique and exciting process. "Lasers remove tattoos by breaking up the pigment colors with a high-intensity light beam. Black tattoo pigment absorbs all laser wavelengths, making it the easiest color to treat. Other colors can only be treated by selected lasers based upon the pigment color" [59].

Skin surfacing is the last laser treatment for skin that is discussed in this section. Skin resurfacing is the precise removal of skin layers one by one with the help of lasers. This will let the skin grow new cells, and in the healing process, the skin looks younger and tighter. This is done for various skin problems such as aging, scars, wrinkles, pigmentation due to sunlight, acne, and other common skin issues [60]. "Selective photothermolysis theory of laser-tissue interaction is used to create thermal destruction of target tissue without unwanted conduction of heat to surrounding structures by selecting the appropriate laser wavelength and pulse duration" [59]. Initially, CO_2 lasers have been used for this treatment, and now more advanced lasers, such as Er:YAG are used to minimize the healing period. In the picture below, we can notice the effect of skin resurfacing to remove scars on a patient after two pulsed dye laser treatments.

Lasers have various applications in dentistry. The applications at a high level can be divided into soft and hard tissue applications. "In the last two decades, there has been an explosion of research studies in laser application. In hard tissue application, the laser is used for caries prevention, bleaching, restorative removal and curing, cavity preparation, dental hypersensitivity, growth modulation, and diagnostic purposes. In contrast, soft tissue application includes wound healing, removal of hyperplastic tissue to uncovering impacted or partially erupted tooth, photodynamic therapy for malignancies, photostimulation of the herpetic lesion" [61]. In addition, the use of lasers in dentistry can be beneficial for children. "Dealing with children for dental issues can be difficult and aggravating because of the complications involved and the fear of parents out of care for their children. Pediatric dentistry is a demanding part of dentistry because of its nature to deal with children from birth through adolescence as well as with their parents' compliance" [62]. Lasers help in making several dental procedures easy with children. According to the American Academy of Pediatric Dentistry (AAPD), lasers help in the dental procedures of both hard and soft tissues. Lasers can provide relief and pain from inflammation caused by ulcers and lesions without pharmacological intervention. They also eliminate the anxiety caused by the noise and vibration of the conventional handheld tools used in dental procedures. In addition, lasers have an analgesic effect on hard tissues eliminating the need for injections. Hence, it makes it easier for dentists to deal with children, and lasers are also known to take relatively less chair time than conventional dental treatments [63].

As discussed earlier, lasers are used for both hard and soft tissue applications. Lasers could help remove overgrown tissues, which otherwise would need a more invasive and painful process. In a classic case of excessive gingival growth, the patient, an eight-year kid, was treated using a laser for a relatively less painful and invasive process [64]. The applications of lasers for eye-related issues are many, and the common eye surgery is LASIK. It is a standard laser procedure to correct vision problems like myopia and hyperopia. Lasik is a widely adopted procedure across the world for correcting vision problems. "If you are short-sighted, the laser flattens your cornea by removing tissue from the center. If you are long-sighted, the laser makes your cornea more curved by removing tissue from the periphery. If you have astigmatism, the laser removes more tissue from some parts of the cornea than others, thus smoothing out the surface to a more regular shape" [65]. The average of Lasik surgery in the U.S was $2,246 per eye in 2019 *vs.* $2,199 per eye in 2018 [66]. The price largely depends on the country where the surgery is performed and the services offered post-surgery.

Besides laser applications for the aforementioned medical procedures, lasers are also used to remove tumors, help prevent blood loss by sealing small blood vessels, and decrease the spread of tumor cells. Different lasers are used for different procedures, based on the type of treatment performed [67]. "As the light beam hits the skin, the skin may either reflect the light away, scatter the light, absorb the light, or let the light pass right through the different layers of the skin. Each layer of the skin uses the light differently" [68].

Finally, from a tribology standpoint, the medical sector is perhaps the most sensitive field affected. Friction takes place between any two moving parts and causes wear, and the human body has several moving parts in relation to each other like teeth, bones, joints, muscles, and other body tissues. These human parts have a naturally low coefficient of friction, and hence our body parts last for several decades. However, wear is a significant problem in the medical sector like automotive and aviation. Though it is not the direct loss of energy due to friction like other sectors, it has more to do with the sensitivity of replacing medical prostheses because of the wear caused due to friction [69]. For example, "A knee prosthesis is a complex medical device used on patients when they can no longer stand the pain due to their arthritis. During their life, the polyethylene prostheses could be affected by wear therefore, preclinical wear tests are necessary to simulate the motion and interaction between the contact materials" [70]. According to a study, surgeons carry out as many as 600,000 knee replacements a year in the United States, and 90 percent of the people have experienced a significant reduction in pain [71]. Bio tribology affects prostheses, medical devices, and surgical instruments, including artificial joints, teeth, dental implants, orthodontic appliances, cardiovascular devices, contact lenses, and surgical instruments [72]. The statistics and data discussed so far are only to highlight the severity of bio tribology. Laser surface texturing is a solution to many of these problems. According to a study, surface texturing decreases wear and increases the load-carrying capacity of the materials used to make hip joint replacements. LST makes dimples that act as a fluid reservoir for the lubricants and entrap the debris from the wear caused by friction. In an experiment performed, it has been found that textured ceramic-on-ceramic artificial hip joints showed a 22% reduced friction and 53% reduced wear compared to a non-textured ceramic-on-ceramic hip joint [73]. This is a perfect example of how laser surface texturing can improve the biotribological properties of certain materials. The effect of tribological features on welded members is also a significant consideration for understanding tribology from a welding standpoint [74 - 95]. From a quality perspective, lasers are minimally invasive, have reduced surgery duration compared to traditional methods, are relatively less painful, and have greater precision because of their ability to narrow down to the target area better, making the surgery more streamlined.

CONCLUDING REMARKS

As discussed at the beginning of this chapter, lasers are used in many industries, and their applications in various fields are only growing with time. From laser printing to Lasik surgery, lasers play a critical role in developing human civilization and helping us progress in technology day by day. Several comparisons have been made in the chapter between conventional machinery and laser-assisted machinery, highlighting how lasers have helped us to be at the forefront of technology, making rapid changes possible, including the improvement of tribological properties, which is a battle that many scientists throughout the world are battling to limit the wear losses from friction between moving parts. From a quality standpoint, lasers have proven one more time to be a clear choice for several applications, with unmatched speed, precision, cost, and effectiveness.

CONSENT FOR PUBLICATION

Not applicable.

CONFLICT OF INTEREST

The author declares no conflict of interest, financial or otherwise.

ACKNOWLEDGEMENTS

The author would like to thank Dr. Muralimohan Cheepu, Super-TIG Welding Co., Ltd., for his support on the discussion of industrial laser tribological applications.

REFERENCES

[1]   Eichler HJ, Eichler J, Lux O. Lasers: basics, advances and applications. Springer 2018.
        [http://dx.doi.org/10.1007/978-3-319-99895-4]

[2]   https://press.uchicago.edu/Misc/Chicago/284158_townes.html#:~:text=Theodore%20Maiman%20mad
        e%20the%20first,rod%20with%20silver%2Dcoated%20surfaces

[3]   Hitz CB, Ewing JJ, Hecht J. An overview of laser technology. 2012: 1-5.

[4]   Jost HP. Economic impact of tribology. Proc 20th Meeting of the Mechanical Failures Prevention
        Group. Gaithersburg, USA. 1974.

[5]   Svelto O. Properties of laser beams. InPrinciples of Lasers. Boston, MA: Springer 2010; pp. 475-504.

[6]   https://ehs.oregonstate.edu/laser/training/laser-types-and-classification

[7]   https://www.onlinelibrary.wiley.com/doi/pdf/10.1002/latj.201870307

[8]   Arnold CB, Serra P, Piqué A. Laser direct-write techniques for printing of complex materials. MRS
        Bull 2007; 32(1): 23-31.
        [http://dx.doi.org/10.1557/mrs2007.11]

[9]   Serra P, Piqué A. Introduction to laser-induced transfer and other associated processes. Wiley-VCH
        2018.
        [http://dx.doi.org/10.1002/9783527805105.ch1]

[10]   https://pdfs.semanticscholar.org/f54d/2d6a15f69c513febd156986a4b2bcab8ffec.pdf

[11]   https://www.fujixerox.com/eng/company/technology/mechanism/s_5process.html

[12]   https://www.ldproducts.com/blog/pros-cons-of-inkjet-and-laser-printers/

[13]   Barnatt C. 3D Printing. Wroclaw, Poland: ExplainingTheFuturecom 2014.

[14]   Ma XL. Research on application of SLA technology in the 3D printing technology. InApplied mechanics and materials. Trans Tech Publications Ltd. 2013; Vol. 401: pp. 938-41.

[15]   https://www.cs.cmu.edu/~rapidproto/students.98/master/project2/applications.html

[16]   https://amfg.ai/2020/01/21/the-evolution-of-sls-new-technologies-materials-and-applications/

[17]   https://nvlpubs.nist.gov/nistpubs/SpecialPublications/NIST.SP.1176.pdf

[18]   Dahotre NB, Harimkar S. Laser fabrication and machining of materials. Springer Science & Business Media 2008.

[19]   https://www.micronlaser.shop/blogs/news/six-reasons-to-pick-laser-over-mechanical-drilling

[20]   Shihomatsu A, Button ST, Silva IB. Tribological behavior of laser textured hot stamping dies. Adv Tribol 2016.
[http://dx.doi.org/10.1155/2016/8106410]

[21]   Etsion I. State of the art in laser surface texturing. J Tribol 2005; 127(1): 248-53.
[http://dx.doi.org/10.1115/1.1828070]

[22]   Romano V, Weber HP, Dumitru G, *et al.* Laser surface microstructuring to improve tribological systems. InLaser Processing of Advanced Materials and Laser Microtechnologies 2003; 5121: 199-211.
[http://dx.doi.org/10.1117/12.515572]

[23]   https://corporate.walmart.com/newsroom/heritage/20141105/6-ways-walmart-made-its-mark-in-retail-history

[24]   https://www.rfgen.com/blog/barcode-scanners-used-by-amazon-to-manage-distributio--center-operations/

[25]   https://www.bbc.co.uk/newsround/47441962

[26]   https://www.consumerreports.org/streaming-music-services/best-music-streaming-service-for-you/

[27]   Ion J. Laser processing of engineering materials: principles, procedure and industrial application. Elsevier 2005.

[28]   https://www.industrial-lasers.com/welding/article/16485079/remote-laser-welding-boosts-production-of-new-ford-mustang

[29]   https://www.sme.org/technologies/articles/2019/october/traditional-versus-laser-welding/

[30]   Thyagarajan K, Ghatak A. Lasers: fundamentals and applications. Springer Science & Business Media 2010.

[31]   https://www.magna.com/company/newsroom/releases/release/2017/04/12/news-release---mgna-receives-gm-supplier-of-the-year-innovation-award

[32]   https://www.autonews.com/article/20160314/OEM01/303149975/honda-laser-cutting-saves--ime-money

[33]   https://www.osram.com/am/specials/trends-in-automotive-lighting/laser-light-new-healight-technology/index.jsp

[34]   https://www.bmw.com/en/innovation/dr-hanafi-and-the-bmw-laserlight.html

[35]   https://www.audi-technology-portal.de/en/electrics-electronics/lighting-technology/matrix--aser-technology1

[36]   https://www.wired.com/story/tesla-use-lasers-clean-glass/

[37]   Etsion I, Sher E. Improving fuel efficiency with laser surface textured piston rings. Tribol Int 2009; 42(4): 542-7.
[http://dx.doi.org/10.1016/j.triboint.2008.02.015]

[38] Ronen A, Etsion I, Kligerman Y. Friction-reducing surface-texturing in reciprocating automotive components. Tribol Trans 2001; 44(3): 359-66.
[http://dx.doi.org/10.1080/10402000108982468]

[39] https://www.aerodefensetech.com/component/content/article/adt/insiders/amm/stories/35490

[40] Katayama S. Understanding and improving process control in pulsed and continuous wave laser welding. InAdvances in Laser Materials Processing. Woodhead Publishing 2018; pp. 153-83.
[http://dx.doi.org/10.1016/B978-0-08-101252-9.00007-8]

[41] https://www.sme.org/fiber-laser-welding-as-a-change-agent

[42] https://www.materials.manchester.ac.uk/research/impact/laser-peening/

[43] https://aerospace.org/article/advancing-laser-communications-technology

[44] https://www.bcsatellite.net/blog/aerospace-advancing-laser-communication-in-space/

[45] https://www.afcea.org/content/harnessing-photons-communications

[46] https://www.militaryaerospace.com/communications/article/16709142/nasa-dod-take-the-next-s-ep-in-laser-communications

[47] https://www.ball.com/aerospace/markets-capabilities/capabilities/electronic-warfare/laser-technologies

[48] https://techxplore.com/news/2015-08-boeing-weapon-drones--aser-welding.html#:~:text=Boeing%20has%20a%20compact%20laser,many%20hundreds%20of%20meter s%20away

[49] Voevodin AA, Zabinski JS. Laser Surface Processing of "Chameleon" Coatings for Aerospace Tribology. InWorld Tribology Congress 2005; 42029: 919-20.
[http://dx.doi.org/10.1115/WTC2005-63489]

[50] Holmberg K, Erdemir A. Influence of tribology on global energy consumption, costs and emissions. Friction 2017; 5(3): 263-84.
[http://dx.doi.org/10.1007/s40544-017-0183-5]

[51] Gaikwad A. Effect of Laser Texturing and Varying Dimple Density on Tribological Properties of Titanium-Ceramic Contact. Rochester Institute of Technology, 2020.

[52] Goldberg DJ, Ed. Laser hair removal. CRC Press 2019.

[53] https://www.surgery.org/sites/default/files/Aesthetic-Society_Stats2019Book_FINAL.pdf

[54] https://mariehayagmd.com/blog/cost-of-laser-hair-removal-treatment-in-the-usa/

[55] https://www.healthline.com/health/laser-treatment-for-hair-loss#does-it-work

[56] https://pubmed.ncbi.nlm.nih.gov/19366270/

[57] https://kiierr.com/what-is-laser-hair-growth-and-how-does-it-work/

[58] Nouri K, Ed. Lasers in dermatology and medicine: dermatologic applications. Springer 2018.
[http://dx.doi.org/10.1007/978-3-319-76220-3]

[59] https://www.webmd.com/skin-problems-and-treatments/laser-tattoo-removal#1

[60] https://www.webmd.com/beauty/laser-skin-resurfacing#1

[61] https://www.ncbi.nlm.nih.gov/pmc/articles/PMC3700144/

[62] Coluzzi DJ, Parker SP, Eds. Lasers in dentistry—current concepts. Springer 2017.
[http://dx.doi.org/10.1007/978-3-319-51944-9]

[63] https://www.aapd.org/research/oral-health-policies--recommendations/use-of-lasers-for-p-diatric-dental-patients/

[64]   Kumar G, Rehman F, Chaturvedy V. Soft tissue applications of Er, Cr: YSGG laser in pediatric dentistry. Int J Clin Pediatr Dent 2017; 10(2): 188-92.
[http://dx.doi.org/10.5005/jp-journals-10005-1432] [PMID: 28890621]

[65]   Sutton G, Lawless M. "Lasik and Asla," in Naked eye, A K A Publishing Pty Ltd 2013.

[66]   https://www.allaboutvision.com/visionsurgery/cost.htm

[67]   https://www.hopkinsmedicine.org/health/treatment-tests-and-therapies/laser-surgery-overview

[68]   https://stanfordhealthcare.org/medical-treatments/l/laser/types/laser-surgery.html

[69]   Valigi MC, Logozzo S, Affatato S. New challenges in tribology: Wear assessment using 3D optical scanners. Materials (Basel) 2017; 10(5): 548.
[http://dx.doi.org/10.3390/ma10050548] [PMID: 28772905]

[70]   Affatato S, Leardini W, Rocchi M, Toni A, Viceconti M. Investigation on wear of knee prostheses under fixed kinematic conditions. Artif Organs 2008; 32(1): 13-8.
[PMID: 18181798]

[71]   https://www.healthline.com/health/total-knee-replacement-surgery/outcomes-statistics-success-rate#positive-outcomes

[72]   Jin ZM, Zheng J, Li W, Zhou ZR. Tribology of medical devices. Biosurf Biotribol 2016; 2(4): 173-92.
[http://dx.doi.org/10.1016/j.bsbt.2016.12.001]

[73]   Roy T, Choudhury D, Ghosh S, Mamat AB, Pingguan-Murphy B. Improved friction and wear performance of micro dimpled ceramic-on-ceramic interface for hip joint arthroplasty. Ceram Int 2015; 41(1): 681-90.
[http://dx.doi.org/10.1016/j.ceramint.2014.08.123]

[74]   Krishnaja D, Cheepu M, Venkateswarlu D. A review of research progress on dissimilar laser weld-brazing of automotive applications. InIOP Conference Series: Materials Science and Engineering 2018; 330(1): 012073. IOP Publishing.
[http://dx.doi.org/10.1088/1757-899X/330/1/012073]

[75]   Srinivas B, Krishna NM, Cheepu M, Sivaprasad K, Muthupandi V. Studies on post weld heat treatment of dissimilar aluminum alloys by laser beam welding technique. InIOP Conference Series: Materials Science and Engineering 2018; 330(1): 012079. IOP Publishing.
[http://dx.doi.org/10.1088/1757-899X/330/1/012079]

[76]   Anuradha M, Das VC, Venkateswarlu D, Cheepu M. Parameter optimization for laser welding of high strength dissimilar materials. InMaterials Science Forum. Trans Tech Publications Ltd. 2019; 969: pp. 558-64.

[77]   Srinivas B, Cheepu M, Sivaprasad K, Muthupandi V. Effect of gaussian beam on microstructural and mechanical properties of dissimilarlaser welding ofAA5083 and AA6061 alloys. InIOP Conference Series: Materials Science and Engineering 2018 Mar 1 330(1): 012066. IOP Publishing.
[http://dx.doi.org/10.1088/1757-899X/330/1/012066]

[78]   Cheepu M, Venkateswarlu D, Rao PN, Kumaran SS, Srinivasan N. Effect of process parameters and heat input on weld bead geometry of laser welded titanium Ti-6Al-4V alloy In Materials Science Forum 2019; 969: 613-8. Trans Tech Publications Ltd.

[79]   Cheepu M, Venkateswarlu D, Rao PN, Kumaran SS, Srinivasan N. Optimization of process parameters using surface response methodology for laser welding of titanium alloy InMaterials Science Forum. Trans Tech Publications Ltd. 2019; Vol. 969: pp. 539-45.

[80]   Cheepu M, Srinivas B, Abhishek N, *et al.* Dissimilar joining of stainless steel and 5083 aluminum alloy sheets by gas tungsten arc welding-brazing process. InIOP Conference Series: Materials Science and Engineering. 330: 012048.
[http://dx.doi.org/10.1088/1757-899X/330/1/012048]

[81]   Cheepu M, Che WS. Friction welding of titanium to stainless steel using Al interlayer. Trans Indian Inst Met 2019; 72(6): 1563-8.
[http://dx.doi.org/10.1007/s12666-019-01655-7]

[82]   Cheepu M, Venkateswarlu D, Rao PN, Muthupandi V, Sivaprasad K, Che WS. Microstructure characterization of superalloy 718 during dissimilar rotary friction welding. InMaterials Science Forum. Trans Tech Publications Ltd. 2019; 969: pp. 211-7.

[83]   Anuradha M, Das VC, Susila P, Cheepu M, Venkateswarlu D. Microstructure and Mechanical Properties for the Dissimilar Joining of Inconel 718 Alloy to High-Strength Steel by TIG Welding. Trans Indian Inst Met 2020; 11: 1-5.
[http://dx.doi.org/10.1007/s12666-020-01925-9]

[84]   Cheepu M, Che WS. Effect of burn-off length on the properties of friction welded dissimilar steel bars. Journal of Welding and Joining 2019; 37(1): 40-5.
[http://dx.doi.org/10.5781/JWJ.2019.37.1.6]

[85]   Anuradha M, Das VC, Susila P, Cheepu M, Venkateswarlu D. Effect of welding parameters on TIG welding of Inconel 718 to AISI 4140 steel. Trans Indian Inst Met 2020; 11: 1-6.
[http://dx.doi.org/10.1007/s12666-020-01926-8]

[86]   Muralimohan CH, Muthupandi V, Sivaprasad K. Properties of friction welding titanium-stainless steel joints with a nickel interlayer. Procedia Materials Science 2014; 5: 1120-9.
[http://dx.doi.org/10.1016/j.mspro.2014.07.406]

[87]   Muralimohan CH, Ashfaq M, Ashiri R, Muthupandi V, Sivaprasad K. Analysis and characterization of the role of Ni interlayer in the friction welding of titanium and 304 austenitic stainless steel. Metall Mater Trans, A Phys Metall Mater Sci 2016; 47(1): 347-59.
[http://dx.doi.org/10.1007/s11661-015-3210-z]

[88]   Muralimohan CH, Haribabu S, Reddy YH, Muthupandi V, Sivaprasad K. Evaluation of microstructures and mechanical properties of dissimilar materials by friction welding. Procedia Materials Science 2014; 5: 1107-13.
[http://dx.doi.org/10.1016/j.mspro.2014.07.404]

[89]   Cheepu M, Muthupandi V, Loganathan S. Friction welding of titanium to 304 stainless steel with electroplated nickel interlayer. InMaterials Science Forum. Trans Tech Publications Ltd. 2012; 710: pp. 620-5.

[90]   Cheepu M, Ashfaq M, Muthupandi V. A new approach for using interlayer and analysis of the friction welding of titanium to stainless steel. Trans Indian Inst Met 2017; 70(10): 2591-600.
[http://dx.doi.org/10.1007/s12666-017-1114-x]

[91]   Muralimohan CH, Muthupandi V, Sivaprasad K. The influence of aluminium intermediate layer in dissimilar friction welds. International Conference on Engineering Materials and Processes. 350-7.

[92]   Jeyaprakash N, Yang CH, Duraiselvam M, Sivasankaran S. Comparative study of laser melting and pre-placed Ni–20% Cr alloying over nodular iron surface. Arch Civ Mech Eng 2020; 20(1): 1-2.
[http://dx.doi.org/10.1007/s43452-020-00030-4]

[93]   Cheepu M, Kumar Reddy YA, Indumathi S, Venkateswarlu D. Laser welding of dissimilar alloys between high tensile steel and Inconel alloy for high temperature applications. Advances in Materials and Processing Technologies 2020.

[94]   Chandra GR, Venukumar S, Cheepu M. Influence of rotational speed on the dissimilar friction welding of heat-treated aluminum alloys. InIOP Conference Series: Materials Science and Engineering 2020 Dec 1 998(1): 012070. IOP Publishing.

[95]   Cheepu M, Cheepu H, Che WS. Influence of joint interface on mechanical properties in dissimilar friction welds. Advances in Materials and Processing Technologies 2020; 1-3.

CHAPTER 5

Laser Surface Hardening

D. Raj Kumar[1,*], **T. Prabakaran**[2] and **D. Gunasekar**[3]

[1] *Department of Mechanical Engineering, MAM School of Engineering, Trichy - 620026, Tamil Nadu, India*

[2] *Department of Mechanical Engineering, MIET Engineering College, Trichy - 620007, Tamil Nadu, India*

[3] *Department of Mechanical Engineering, Jayaram College of Engineering and Technology, Trichy- 621014, Tamil Nadu, India*

Abstract: The basic materials have been potentially applied to the automobile, marine, aerospace and biomedical sectors. Surface modifications are required to increase tribological behaviors and mechanical properties of basic material surfaces using lasers. The material surface properties are improved through laser without altering the bulk. The surface modifications on materials have been focused on by many researchers due to their needs. However, the laser surface modifications on materials are a suitable method for improving tribological and mechanical properties of the surface due to the number of features and economy. In this chapter, laser surface hardening has been analyzed based on the microstructure, wear, coefficient of friction, microhardness, surface roughness, worn-out surface, and tensile strength. Therefore, laser surface hardening is recognized as an important topic based on surface engineering and metallurgical limitations.

Keywords: CO_2 laser, Coefficient of friction, Laser surface hardening, Mechanical properties, Microhardness, Nd:YAG, Surface roughness, Tensile, Tribological properties, Wear, Worn out surface.

INTRODUCTION

The mechanical component performance is enhanced by several methods, such as mechanical impact (shot peening, water peening, burnishing and hammer), thermochemical impact (case hardening, carbo nitriding, nitriding, boronizing and chromizing), thermal impact (flame hardening, induction hardening, laser hardening and surface layer remelting), thermomechanical impact (friction hardening) and surface layer hardening by machining. Among the various

---

* **Correspondence author D. Raj kumar:** Department of Mechanical Engineering, MAM School of Engineering, Trichy - 620026, Tamil Nadu, India; E-mail: profdrajkumar@gmail.com

Jeyaprakash Natarajan and Che-Hua Yang (Eds.)
All rights reserved-© 2021 Bentham Science Publishers

methods, laser surface treatment is almost applied to all metallic materials in the industrial sectors for improving the mechanical behavior and tribological properties of the surface [1]. This is because of minimum heat input, less distortion, no further finishing work required, low hardness stress, lesser risk of cracking and no media required for quenching [2]. Laser is the main source of heat energy and commercially available in Neodymium Yttrium-Aluminiu- -Garnet (Nd:YAG) solid type, (Carbon-di-Oxide) CO_2 laser and diode lasers, which are operated in pulsed or continuous power. These lasers are widely used in surface treatment for improving tribological and mechanical properties [3]. In industries, most of the machine components are faced with worn-out surfaces. Hence, improvements are required in the surface properties of the part. Therefore, a laser source is required. The selections of laser sources also affect the properties of the surface. This is due to the different laser sources having different wavelengths. Also, surface properties are related to laser source wavelength, energy input, and absorptivity of the material and polished metal surface. The laser surface alteration is the process in which the process can vary the thermal energy from the laser source to work material.

In laser hardening, the heat energy can be split into two portions in which one portion of heat energy is absorbed by using the material surface. The remaining portion of heat energy is reflected from the surface. Moreover, to improve the wear resistance properties of the surface, laser surface hardening is selected. During the laser hardening with a short interval time, the thermal energy of the laser source directly affected the material surface without affecting base metal properties. The major benefits of laser surface hardening are increased hardness, reduced friction, reduced wear, increased fatigue life, enhanced strength and the formation of unique geometrical wear. The heat inputs to metal surfaces are used for forming fine-grained microstructure of laser processed samples. The lower crack formations are observed owing to the self-quenching. Nd:YAG laser is selected for the hardening of steel surface because of the short laser wavelength of 1.064 μm and absorptivity of the surface is increased through short wavelength. The higher wavelength of CO_2 laser (10.6 μm) is used to lower coupling interaction with metallic substrates. Therefore, coating or painting is applied to the metal surface to increase the absorption rate. Moreover, the painting or coating is highly influenced by environmental conditions. Therefore, the performance of the Nd:YAG laser is superior to the CO_2 laser. This is due to the short wavelength of the Nd:YAG laser. The effect of short-wavelength on surface material is used to increase the absorbing rate of the surface. The fiber cable is used to transfer heat energy from the Nd:YAG laser and the CO_2 laser is inadequate. To eliminate atmospheric contamination, inert gases are used. Also, for reducing the wavelength of the laser, an excimer laser is developed for micromachining on surgical components.

In this book chapter, surface hardening by laser has been discussed for tribological and mechanical properties, such as microstructure, wear, Coefficient of Friction (CoF), microhardness, surface roughness, worn-out surface and tensile strength. Therefore, laser surface hardening is focused as an important topic for improving surface properties.

LASER SURFACE HARDENING

Laser Surface Hardening (LSH) is expressed as the laser beam heat energy that is directly processed on the work material surface for improving the tribological properties without affecting the bulk. It is one of the methods of laser surface treatment [4]. The schematic of LSH is shown in Fig. (**1**). To improve the hard martensite surface, quenching is required. Therefore, the surface properties are improved by laser and the desired tribological properties are achieved. The mechanical components, such as exhaust valves, gears, gear teeth, cylinder liners, camshafts, and valve guides normally involve the higher stresses on the surface during the processing. To improve surface properties, LSH is needed for improving tribological properties. The high-stressed components are mostly made up of medium-carbon steel, cast iron and die steel. The electronic parts, mass-production industries and automobile components require the laser hardening process on the part surfaces. The surface properties are controlled using scanning speed, beam shape, power, scanning speed and material surface conditions. The surface properties of the I-section rail are also increased through the LSH [5]. The LSH is required on the precision components for improving hardness by laser. The high cooling effect produces a high hardness surface. The water, oil and air are used as quenching mediums for phase transformation of the surface. Thereby, laser surface hardening provides better performance than flame and induction hardening. Therefore, a comparative wear rate and hardness for steel are performed by laser and conventional quenching. The conventional quenching temperature of 1198 K is maintained at 4.5 h, whereas the laser quenching temperature of 523 K is maintained at 4 h. The linear speed of 168 mm/s, air as medium, the spot diameter of 3.5 mm and 10 kW CW diode laser are used. The 600 $HV_{0.1}$ and 625 $HV_{0.1}$ hardness are observed in the laser and conventional quenched surfaces, respectively. The laser quenched process has a lesser wear rate produced compared to conventional quenched [6]. To provide better wear resistance and hardness than conventional treatment, a high-frequency quenching method is applied to 40CrNiMoA steel. A CW CO_2 laser, 60-degree incident angle, 35 mm/s traverse speed and 1400 W laser power, black organic absorbent coating, 10 mm defocusing distance and 0.9 m^3/h gas flow rate are employed in the laser treatment. It was found that better hardness & wear resistance is observed in the high-frequency quenching method [7].

A comparison is made between the different methods of laser surface hardening over the different types of structural, gray cast irons, tool steels, special maraging steel and hardening of nodular graphite. The method such as the selection of absorbent, laser surface hardening, laser surface remelting, laser shock hardening is considered in this work. The analysis found that laser surface hardening is better to improve surface quality compared to other methods. The type of absorbent, different degrees of overlapping, different modes of laser-beam guidance and different energy input decide the quality of the surface. It is highly affected by microstructure, microhardness and surface roughness [8]. The main reasons for selecting the LSH are:

- Low energy is used compared to other surface heat-treatment.
- The energy input is varied by changing the travel speed of work material, different levels of defocus, selection of lenses and mirrors of different shapes.
- The self-quenching.
- The small parts are also heat treated.

The demerits for laser surface hardening are:

- High capital cost
- Need of surface preparation in a special process
- Required protection over radiation
- Need skilled operators

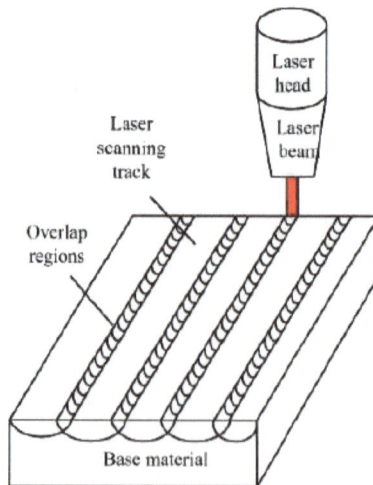

Fig. (1). Schematic of laser surface hardening.

The laser hardening is also required in the high stress acting on the mechanical component surface. Hence, the wear resistance of the load-bearing component is necessarily increased by laser. The results found that the laser treatment has effectively improved the wear resistance properties of the material by changing the fine microstructure [9]. The microstructure formation is highly dependent upon laser parameters, which is shown in Fig. (**2**).

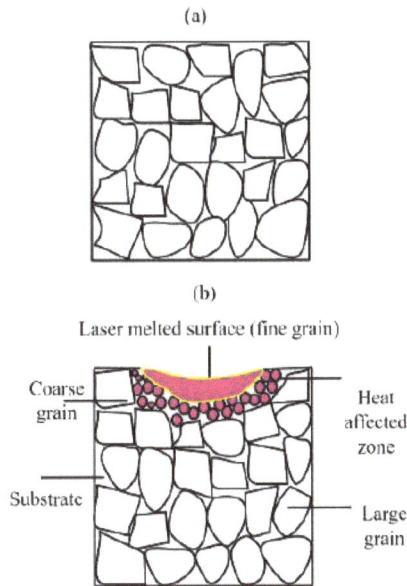

Fig. (2). Schematic of as received tool steel microstructure (**a**), laser surface hardened with modified structure (**b**).

In Fig. (**2**), in the parent substrate surface, it was found that the coarse and unequal size equiaxed grains are presented in the substrate and heat-affected zone. The surface hardened by laser is to enhance material properties by changing microstructure from large grain into a fine grain. The depth of hardening is varied by increasing the grain size from fine to coarse. During the LSH process, the curved shape surface is formed in the top surface work material. The lower scanning speed generates more evaporation in the top surface. To obtain the desired hardness on the surface, 1.5 kW power, 0.18 s interaction time, 3 mm beam diameter and 1 m/min scan speed are used. The argon flow rate at 20 L/min is utilized to decrease hardness. It is identified by changing the distance from the top surface to a depth of 200-micron, which is illustrated in Fig. (**3**). Also, the performance of camshaft and crankshaft are increased by using argon gas, 3 mm beam diameter, 1 m/min velocity and 1.5 kW power [10].

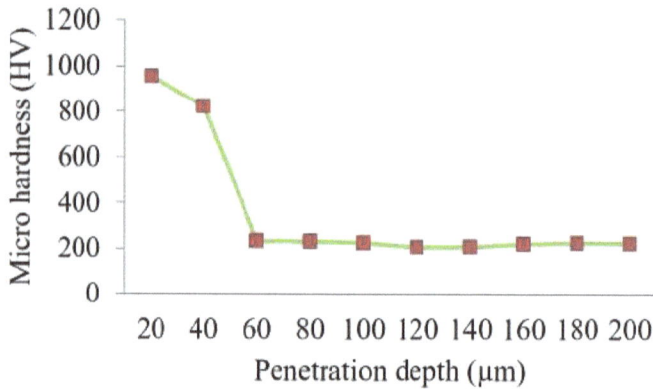

Fig. (3). Microhardness variations from the top surface to substrate through LSH.

MICROSTRUCTURE

The microstructure is a material structure that is visible at the micron level. The microstructures for laser processed materials are used to examine the phase of materials, impurities, defects, grains, and grain boundaries by digital microscope or Scanning Electron Microscope (SEM). The primary goal of laser hardening is used to enhance friction, wear reduction and increased mechanical properties. The service life of the automobile component is increased by the proper selection of laser sources, laser processing variables and absorption coating. These factors decide the microstructural changes on the material and affect the tribological and mechanical properties of the surface [11]. In the case of steel hardening, LSH is applied to steel. The adequate carbon content in metal surfaces allows hardening and forming pearlite microstructure. The laser surface hardened is performed on En18 steel by laser parameters, such as power at 1.5 kW, beam diameter at 3 mm and scan speed at 1 m/min. The macrostructure of laser surface hardened steel surfaces are classified as pearlite & ferrite, partially transformed and highly hardened martensite region. The SEM microstructures of laser surface hardened steel surfaces showed that finer martensite is formed near the surface, the homogenous martensitic is formed towards the middle of the laser-treated area and partially formed in the transformed heat-affected zone [12]. The laser surface hardening of gunmetals is performed by varying numbers of laser shots. The microstructures of gunmetals showed that a small number of scratches, re-solidification metal surface, long, spaghetti-like structure, circular, oval and banana-like structures are observed in the laser processed microstructure.

The particle size is greatly related to the number of laser shots and it is affected by the hardness [13]. The laser surface hardening of AISI 410 martensitic stainless steel is performed by a pulsed Nd:YAG laser varying the power, focal position, scanning speed and pulse width. The influence of warm work and air quenched processing on AISI 410 martensitic stainless steel is also analyzed. The delta ferrite phase in the martensitic base and more ferrite phase is used to increase surface hardness [14]. The microstructure of the 1538MV steel surface is studied by using Conventional Quenching – Tempering (CQT) and Laser Quenching (LQ). The CQT is carried out at an austenitization temperature of 1198 K in 4.5 h, whereas LQ is carried out using a diode laser, a spot diameter of 3.5 mm and linear speed at 168 mm/s. The laser parameters alter the microstructure of the base metal, laser quenched surface and conventional quenching-tempering surface and grain morphology, which is shown in Fig. (**4**). It is clear that the raw sample shows disperse perlitic and ferritic phases and it is shown in Fig. (**4(a)**). The fine grains with homogeneous martensitic phase are formed in the laser quenched portion and it is shown in Fig. (**4(b)**). The non-uniform martensitic phases are shown in Fig. (**4(c)**). Moreover, LQ produces less wear due to fine grains formed in the LQ portion [15].

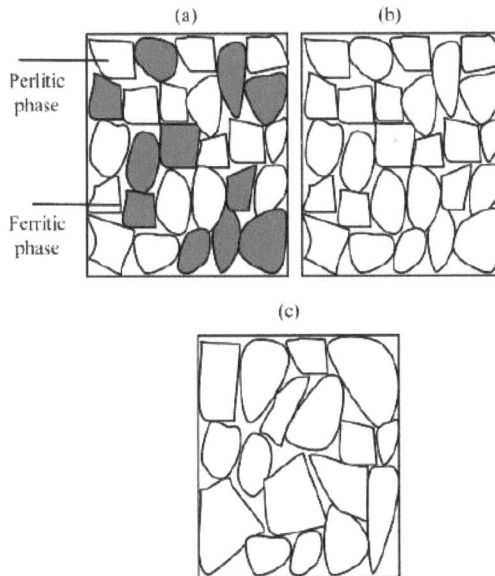

Fig. (4). Grain morphology of the longitudinal section (**a**) perlitic and ferritic phases for base material, (**b**) homogeneous martensitic phase, and (**c**) non-uniform martensitic phase.

The microstructure of Gray Cast Iron (GI) surface is analyzed by varying quench-tempering (LHQT), laser surface hardening (QT) and austempering (A). The three regions are formed in the laser processed sample, such as ledeburite, martensite and tempered martensite. The ledeburite zone is observed in the solid–liquid–solid phase transformation. After the effect of self-quenching on the surface, the martensite phase is observed and it is closer to the heat-affected zone. The graphite flakes are also maintained in the martensitic phase. On increasing the tempering temperature with constant holding time, a tempered martensitic phase is formed [16]. The austempering temperature varies from 232 to 399° C, the ferrite is transformed in the needle-like shape into a feather-like shape and it is shown in Fig. (**5**).

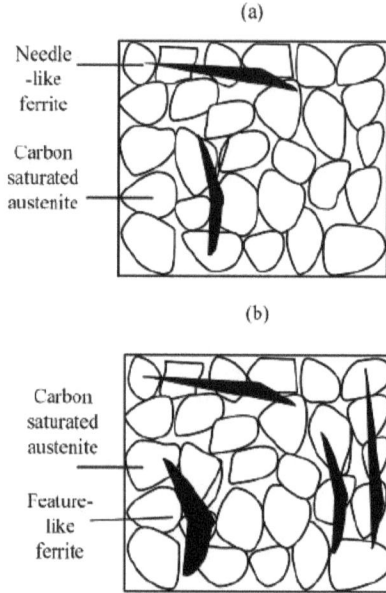

(a)

Needle -like ferrite

Carbon saturated austenite

(b)

Carbon saturated austenite

Feature- like ferrite

Fig. (5). Microstructure of gray cast iron (**a**) 232°C, (**b**) 399°C.

WEAR

Wear is a common phenomenon in any mechanical functional component. The wear can be reduced by a number of methods, such as conventional heat treatment and laser surface treatment [17]. Laser is a better option for improving the surface properties of the work material. Many parameters were used to select the LSH, including power, scanning speed, cooling rate, heating, and peak temperature. These parameters constantly vary the microstructure and surface morphology. Also, the development of high-power diode lasers increases the power density and decreases the cost. Moreover, the wear can be better prevented by selecting suitable laser surface hardening conditions. The basic mechanism for improving wear resistances is by changing the microstructure of the surface without altering bulk material properties. The wear test is mostly processed on pin-on-disc apparatus and it is shown in Fig. (**6**). The wear test is conducted by varying the sliding distance and applied load. The wear can be calculated using a number of terms, such as wear rate and lost volume.

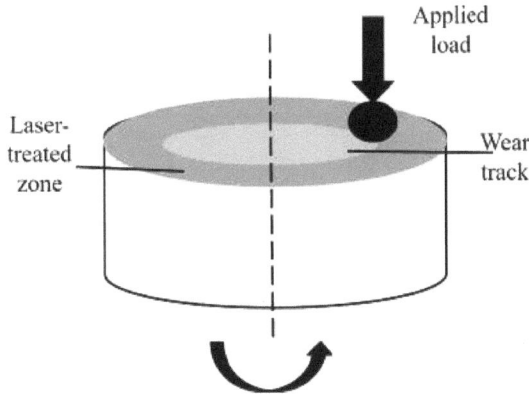

Fig. (6). Schematic for pin-on-disc apparatus.

Wear rate is calculated by the wear scars formed in the steel surface using tungsten carbide balls. Lost volume is calculated by the following equation.

$$V = \frac{\pi b^4}{64R^3} \qquad (1)$$

Here, V is denoted as the lost volume, b is denoted as wear scar radius, and R is denoted as sphere radius in the case of $b \ll R$.

Wear rate is calculated on the basis of Archard law, which is illustrated in the following equation

$$V = KsN \qquad (2)$$

Here, V is denoted as the lost volume; k is denoted as the wear constant; s is denoted as the sliding distance, and N is denoted as the applied load. Hence, the wear rate is calculated by the following equation:

$$K = \frac{\pi b^4}{64Rs\,N} \qquad (3)$$

As per ASTM G99, specific wear rate (K) is evaluated, and the equation is given below:

$$wear\ rate\ =\ \frac{wear\ volume}{load\ \times sliding\ distance}\quad \left[\frac{mm^3}{Nm}\right] \tag{4}$$

The wear resistance properties of twin valves in the internal combustion are increased by laser surface hardening. The 0.5 mm hardened layer depth, 850 Hv hardness and 0.16 mm melted layer depth are observed in the twin valve using the 1300 °C temperature, 600 W power, 2.7 J/mms^2 energy density [9]. The camshafts, axles, spindles, steering levers, connecting rods, and gears surface properties are enhanced through CW CO_2 laser surface hardening. The automobile components are mostly made up of En18 steel. The laser power varying between 1.3 and 1.5 kW, scan velocity of 1 m/min, beam diameter of 3 mm and helium shielding are used for increasing the wear depth and hardness. The results found that the 925 $HV_{0.2}$ hardness and 500-micron wear depth are achieved by using 1.5 kW power, 20 N applied force and 200 mm sliding distance [12]. The CK60 structural steel surface properties are enhanced by CO_2 laser surface hardening. The results observed that the hardened depth varies from 50 to 430 mm in the heat-affected zone, where wear resistance is increased by 38% overlapping ratios [15]. The base material of 1538 MV steel surface properties is improved by using LSH. In order to get better base material properties, conventional quenching – tempering (CQT) and laser quenching (LQ) are performed on the steel. The CQT is conducted at the austenitization temperature of 1198 K in the duration of 4.5 h, whereas LQ is carried out using a diode laser, spot diameter at 3.5 mm and linear speed at 168 mm/s. The graph is made between the specific wear rate and sliding distance, and it is shown in Fig. (7). The results found that the raw steel (RAW) has a higher wear rate than CQT and LQ due to the high ductility. Moreover, LQ produces less wear due to fine grains formed in the LQ portion [16].

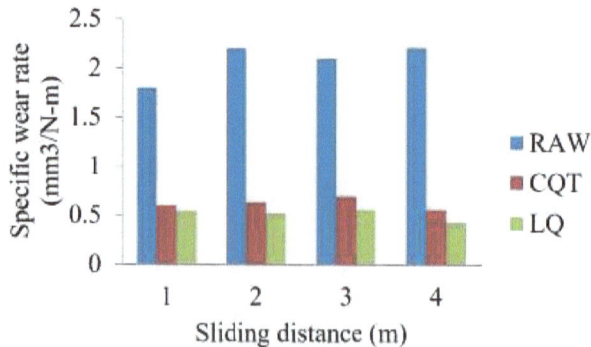

Fig. (7). Wear rate for 1538MV steel.

Gray cast iron (GI) is made up of gears, engine liners, guide rails, camshafts and engines. The mechanical components often face wear problems and reduce service life and performance. Moreover, on increasing the surface properties of GI, quench-tempering (LHQT), laser surface hardening (QT) and austempering (A) are performed on GI. The effect of LHQT, QT and A on wear loss GI is drawn and it is shown in Fig. (**8**). The results found that the untreated GI produces higher mass loss due to the low hardness. The QT has lower mass loss compared to LHQT due to the reducing surface hardness [18].

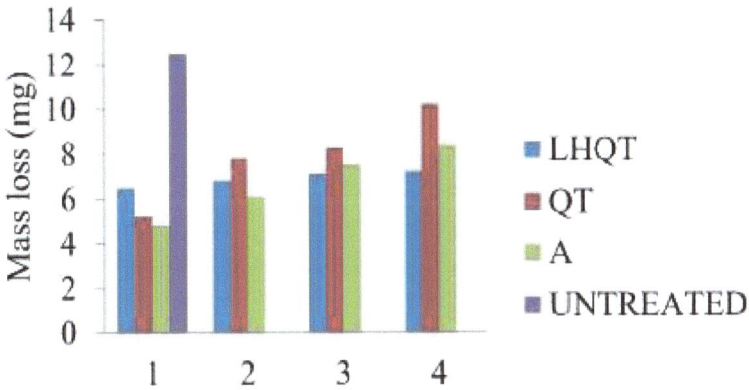

Fig. (8). Mass losses for grey cast iron.

COEFFICIENT OF FRICTION

The Coefficient of Friction (CoF) is the ratio of the force required to move two sliding surfaces over each other. Generally, smaller than 0.1 CoF is considered lubricous material. The CoF depends on the nature of the materials and surface roughness. Usually, the ASTM D1894-14 standards for CoF measurement are used [19]. The schematic of CoF is shown in Fig. (**9**).

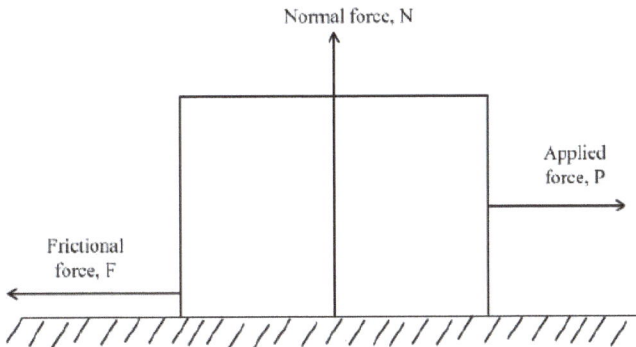

Fig. (9). Schematic for the coefficient of friction.

$$Coefficient\ of\ friction, \mu = \frac{F}{N} \tag{5}$$

Where F is denoted as the frictional force; N is denoted as the normal applied load. The CK60 structural steel frictional properties are enhanced by CO_2 laser surfaces hardening conditions, such as the power of 2.5 kW, the velocity of 3 m/min, the defocused distance of 60 mm and 0%, 9%, 21% and 38% overlapping. The analysis found that the coefficient of friction is reduced by reducing the number of cycles [15]. The base material of 1538MV surface properties is improved by using LSH. For improving the base material properties, conventional quenching – tempering (CQT) and laser quenching (LQ) are performed on the steel. The CQT is performed at an austenitization temperature of 1198 K in the duration of 4.5 h, whereas LQ is carried out using a diode laser, spot diameter at 3.5 mm and linear speed at 168 mm/s. The results found that the LQ samples produced the lowest coefficient of friction [16]. The conventional method of hardening on the thin sectioned and high-precision parts is a difficult process due to the inhomogeneous hardness distribution, large heat-affected zone, surface deformation and overheating. Premature failures are possible in the components. To overcome the conventional problems, LSH is performed on thin-section material. The 100Cr6 steel surface properties are improved by LSH with four methods, such as continuous wave mode of processing (CW), continuous wave mode of processing with fluid beneath the sample (CW-UF), pulsed wave mode of processing (PW), pulsed wave mode of processing with fluid beneath the sample (PW-UF) methods. The results found that the laser surface hardening with PW-UF methods reduces distortion and improves the tribological performance [20]. The effect of power, defocusing distance and speed on CoF SAE 4130 sheets of steel are studied by using LSH conditions. The laser power at 600 W, a speed of 8 mm/s and defocusing at 50 mm is used. The results found that the CoF of the laser processed surfaces is half of the bare parameters [21]. The average CoF and maximum CoF are better for the bare conditions when compared to the laser conditions, and it is shown in Table **1**.

Table 1. CoF for SAE 4130 steels.

Sample	Maximum CoF	Average CoF
Bare	0.51	0.35 ± 0.07
V8	0.28	0.21 ± 0.03
V10	0.28	0.18 ± 0.03

Low alloy and EN25 steel are designed for providing better properties when evaluating conventional carbon steels. The low alloy steels are also potentially applied to high-temperature parts such as gear shaft, hot forging die, connecting rod and piercing tool. In order to improve the surface properties of EN25 low alloy steels above quenched and tempered conditions, LSH is required for critical application components of low alloy steels. The impact of applied load, temperature and sliding distance on CoF of EN25 steel is studied by LSH on steel. The LSH is performed on steel by varying power, and travel speed and using argon gas as shielding gas. The results found that the laser processed material has lower CoF than the received CoF. The CoF of laser hardened EN25 steel is varied from 0.046 to 0.189, whereas as received CoF is varied from 0.33 to 0.49 using 400°C. It is also noticed that the coefficient of friction is increased by increasing applied load and sliding distance. Lesser coefficient of friction is observed at the temperature of 400°C and it is increased by increasing temperature of 600°C [22].

MICROHARDNESS

The hardness is expressed as the resistance of a metal to resist the local plastic deformation in the flat surface by applying load. The hardness of the surface of the component is an important property affecting the tribological performance. The dynamic components, such as rolling bearings, gears, and plain bearings, require higher surface hardness for avoiding high wear loss. The commonly used hardness tests are defined by the type of indent, size, and load applied. The two hardness test methods are used as Brinell hardness and Rockwell hardness. Each method has its unique hardness scales. Hardness is correlated approximately to ultimate tensile strength in the metals. A study is used to investigate the effect of laser process parameters on hardness austempered ductile iron by spot LSH. The power varies from 420 to 1140 W, and the beam diameter varies from 1.25 to 2 mm; 300 ms pulse duration and 7.5 l/min nitrogen gas flow rate are used in the spot LSH. The laser hardened spots are formed by using low powers at 600 W, beam diameter at 1.75 mm and pulse duration at 300 ms. The results observed that the surface hardened varies from 700 to 800 HV in the depth of 150 μm [23]. The hardness comparisons are performed on low-medium and upper austempered ductile iron grades by Nd:YAG laser surface hardening process. The results found that the uniform microhardness of 400 HV is observed in the grades due to the tempered martensite and bainite phases. The mechanical performances of these grades are also approximately equal [24]. The hardness of the S45C medium carbon steel property is improved by using uniform beam technology in the high power Nd: YAG laser with a continuous wave. The parameters, width, focal length, and the length of the optical lens are 40, 195 and 40 mm, respectively. The results of the output found that the hardness and width of the hardened surface are 780 Hv and 22.3 mm, respectively [25]. SAE 4130 steel has been used in oil pipelines, aircraft, automotive gearboxes and landing gears. The steel possesses low to medium hardenability. To increase hardness above this material level, the effect of laser process parameters, power, defocusing distance and scanning speed on hardness SAE 4130 sheets of steel is studied by varying LSH conditions. The 600 W laser power, a speed of 8 or 10 mm/s and defocusing of 50 mm are used in the LSH. The graph is made between the hardness and distance from the surface and it is shown in Fig. (10).

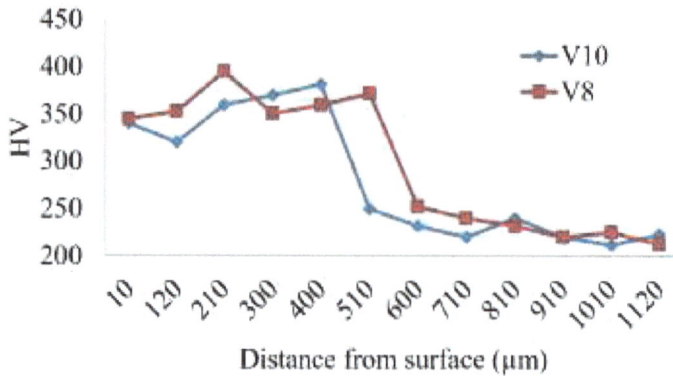

Fig. (10). Hardness over distance from surface for SAE 4130 steel.

It is shown that the hardness varies from 320 to 400 HV in the laser processed sample, which is higher than the base metal. The hardness is increased by about 80% from the base metal. V8 is represented as a partially transformed region, whereas V10 is represented as a hardened layer in the tempered region by laser [21]. The LSH is carried out on EN25 steel to evaluate microhardness for the laser hardened. The untreated specimens are hardened by varying load ranging from 10 to 40 N, sliding distances from 1000 to 3000 m, and varying elevated temperatures ranging from 200°C to 200°C at 0.15 m/s sliding speed. The laser spot diameter size at 1.55 mm, the focal position at 10 mm and the lens focal length at 300 mm are used in the Nd:YAG laser. The effects of depth below the surface and power on microhardness are graphed, and it is shown in Fig. (11).

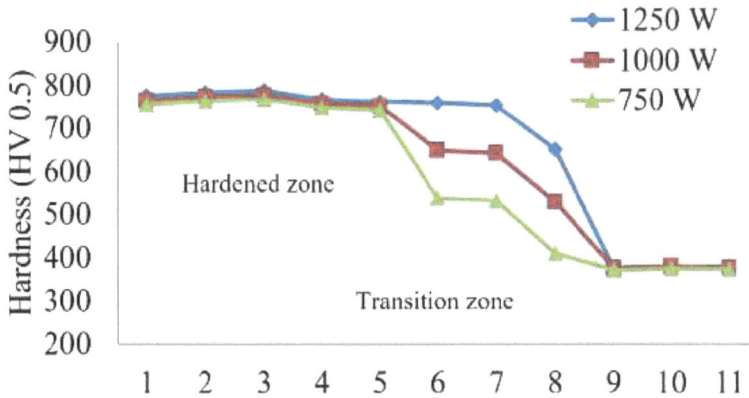

Fig. (11). Hardness over distance from the surface of EN25 steel surface roughness.

On raising the power, hardness depth is increased. The hardness difference between the laser hardened zone and the transition zone is low. The maximum microhardness of 780 ±10 HV is achieved in the laser hardened surface [26].

SURFACE ROUGHNESS

Surface roughness (SR) is expressed as the number of deviations in the machined surface. More deviation and less deviation of the machined surface represent the rough and smooth surfaces, respectively. The surface roughness can be measured by both the contact method and the non-contact method. A roughness study is performed to investigate the effect of laser process parameters, ground state, laser hardened and laser melted portion on austempered ductile iron by spot LSH. The parameters such as power varying from 420 to 1140 W, beam diameter varying from 1.25 to 2 mm, 300 ms pulse duration and 7.5 l/min nitrogen gas flow rate are used in the spot LSH. The laser hardened spots are formed at low powers of 600–720 W, beam diameter of 1.75 mm and pulse duration of 300 ms. It was observed that the LSH spot has produced slight distortion and lower surface roughness compared to laser melted portion surface roughness. Also, the ground state has produced lower surface roughness compared to the laser hardened surface roughness [23]. The incident angle of the laser parameter plays a major role on the flat and curved surface of the crankshaft fillet. To evaluate the surface roughness of flat and curved surface energy absorption, laser surface hardening and energy absorption are performed. The 2 kW fiber laser, the beam shape of 9 mm diameter, the zero focal position, the feeding rate of 10 mm/s and no shielding gas are used in the LSH.

The results found that no difference in the surface roughness between the flat and curved surfaces is observed [27]. The effect of LSH parameters on the surface roughness of gray cast iron and ductile cast iron is investigated. The laser conditions, such as a power of 1200 W, beam size of 6.35 mm and scanning rate of 8.47 mm/s are used for hardening gray cast iron, whereas a scan rate of 12.70 mm/s is used for ductile cast iron. The results found that the cast iron with an untreated machined surface has produced the surface roughness of 45 – 55 μin, whereas laser heat-treated surface has produced the surface roughness of 45 – 60 μ in [28]. The influence of laser heat treatment (LHT), multi-pin ultrasonic impact treatment (UIT), UIT + LHT and LHT + UIT processes on medium-carbon steel AISI 1045 is performed by laser. The laser beam spot diameter of 1 mm, the specimen feed rate of 90 mm/min, duration of the laser action on the treated area of 0.66 s are used in the LHT process. The vibration frequency of 21.6 kHz and amplitude of 18 μm is used in the UIT. The frequency of vibration-induced is 2.0 ± 1.0 kHz. The results found that the combined LHT + UIT treatment produced a 0.25 μm surface roughness and the highest surface hardness is observed in the hardness of 11,000 HV0.05 [29]. The surface roughness of the piston ring is enhanced by varying LSH conditions, such as power varying from 1.0 to 4.5 kW, speed varying from 1 to 13 m/min, a focal length of 311 mm and an incident angle of 90°. The results identified that the surface roughness is observed in the range from 1.6 μm to 12.6 μm by using heat ranges between 30 and 45 J/mm. Also, the surface roughness is decreased by increasing travel speed and decreasing laser power. The lower travel speed and higher laser power are used to obtain minimum surface roughness [30].

WORN OUT SURFACE

The LSM is conducted on metallic material. The wear test is carried out on laser process samples by pin-on-disc tribometer. The wear tracks are formed in the laser processed samples. The worn-out surface of wear tracks is analyzed based on the formation of defects. The defects such as smearing wear, crack and ploughing wear are normally observed in the worn-out surface. These defects are formed based on the laser treatment method, absorbent, tempering, quenching and austempering temperatures. The impacts of quench-tempering and laser hardening on worn tracks of cast iron are studied. The various tempering temperatures and constant holding temperature is used. The impact of quench-tempered gray cast iron analysis is conducted on worn-out areas. It is found that the worn areas in the laser hardened zone are quite smooth due to the polishing by the upper ceramic ball specimen. Also, cracks are formed in the substrate region and formed around the tips of graphite flakes. This is due to the high-stress concentration and developed along with the graphite flakes. The small pits and spalls are detected in the worn-out surface due to the linkage of cracks on the surface and subsurface [18]. The influence of laser surface hardening process parameters on microstructures and wear behavior of ductile iron is studied by varying austempering temperatures of 232°C, 288°C and 398°C with different laser gaps of 1.5 mm, 3 mm and 4 mm. The laser processed surface has ledeburite and tempered bainite microstructures. Also, smearing wear and ploughing wear are observed in the tempered bainite zone [31]. The LSH is carried out on EN25 steel for evaluating worn-out surfaces by varying different loads, sliding distances, and varying elevated temperatures with constant sliding speed. The laser spot diameter size of 1.55 mm, the focal position of defocusing to 10 mm in the negative direction, and the lens focal length of 300 mm are used in the Nd:YAG laser. The results found that the localized fracture of oxide transfer layers is formed at 600°C in the laser processed sample. The cracks and adhesion are also formed. The localized fracture of oxide transfer layers is denoted as exfoliation. This mode of failure is more severe than delamination on the surface [26].

TENSILE STRENGTH

The tensile strength of laser surface hardened is defined as the maximum load that material is supported without fracture when being stretched, divided by the original cross-sectional area of the material. It is also an important property of the laser surface hardened materials. The tensile properties, ultimate tensile strength, yield strength and percentage of elongations are presented based on the effect of laser process parameters on the surface. The temperature and surface hardening of low carbon thin steel sheets are analyzed by using a Yb-fiber laser. The results found that the yield strength and tensile strength are increased from 46-80% and 22-45%, respectively, due to laser surface hardening. Moreover, the percentage elongation of 30% was reduced by increasing the tensile strength of the laser processed sample [32]. The influence of laser surface hardened layers on mechanical properties of re-engineered low carbon steel sheets is studied by laser. The fiber-coupled diode laser, 600-1000 W power, 12-32 mm/s scanning speed, 4 mm x 4 mm beam spot, cold-rolled annealed surface condition, 950°C - 1000 °C transformation hardening temperature and 12.5% overlap multi-track area hardening are used in this LSH. The results found that the tensile property of the layered steel sheet is improved in yield strength in the range of 25-40% and ultimate tensile strength of 20% due to the laser surface hardening [33]. To improve the tensile properties of carbon steel, laser surface hardening is performed by using laser parameters, such as Nd:YAG laser, 14 kW power, 0.1-3.0 mm spot diameter, 0.5-10 ms pulse duration and 0-20 HZ pulse frequency. The results observed that the 412 MPa yield strength, 665MPa ultimate strength is observed in the sample due to the LSH [34].

CONCLUDING REMARKS

The basic materials are seldom applied to industrial and scientific components because of their limited properties. Hence, to improve the basic surface properties, several improvement methods are presently available, such as mechanical impact, thermochemical impact, thermal impact and thermomechanical impact. Laser surface hardening is a potential method and is widely applied to all metallic materials for improving the tribological and mechanical properties due to the precise heat input, low distortion, less hardness stress and minimum risk of crack propagation. The properties, microstructure, wear, coefficient of friction, microhardness, surface roughness, worn-out surface and tensile strength has mainly been focused based on the various laser surface hardening treatment, quenching and austempering temperature. The laser surface hardening is applied to high stressed components of medium-carbon steel, cast iron, die steel, electronic parts, mass-production industries and automobile components. The surface properties are controlled by varying scanning speed, beam shape, power, scanning speed and modifying material surface conditions. Hence, laser-based surface hardening is adopted for improving the surface properties of the components.

CONSENT FOR PUBLICATION

Not applicable.

CONFLICT OF INTEREST

The author declares no conflict of interest, financial or otherwise.

ACKNOWLEDGEMENTS

I would like to thank the MAM School of Engineering, Tiruchirappalli, India, for providing scientific details and experimental work. I wish to extend my special thanks to Editor Dr. N. Jeyaprakash for his valuable guidance and support in completing the chapter.

REFERENCES

[1]   Jeyaprakash N, Duraiselvam M, Aditya SV. Numerical modeling of WC-12% Co laser alloyed cast iron in high temperature sliding wear condition using response surface methodology. Surf Rev Lett 2018; 25(07): 1950009.
[http://dx.doi.org/10.1142/S0218625X19500094]

[2]   Dinesh Babu P, Balasubramanian KR, Buvanashekaran G. Laser surface hardening: a review. Int J Surface Sci Eng 2011; 5(2-3): 131-51.
[http://dx.doi.org/10.1504/IJSURFSE.2011.041398]

[3]   Davis JR, Ed. Surface hardening of steels: understanding the basics. ASM international 2002.

[4]   Jeyaprakash N, Yang CH, Duraiselvam M, Prabu G, Tseng SP, Kumar DR. Investigation of high temperature wear performance on laser processed nodular iron using optimization technique. Results Phys 2019; 15: 102585.
[http://dx.doi.org/10.1016/j.rinp.2019.102585]

[5]   Jeyaprakash N, Yang CH, Kumar DR. Laser Surface Modification of Materials. InLaser Ablation. IntechOpen 2020.

[6]   Carrera-Espinoza R, Rojo Valerio A, del Prado Villasana J, *et al.* Surface Laser Quenching as an Alternative Method for Conventional Quenching and Tempering Treatment of 1538 MV Steel. Adv Mater Sci Eng 2020; 2020.

[7]   Liu Q, Song Y, Yang Y, Xu G, Zhao Z. On the laser quenching of the groove of the piston head in large diesel engines. J Mater Eng Perform 1998; 7(3): 402-6.
[http://dx.doi.org/10.1361/105994998770347855]

[8]   Grum J. Comparison of different techniques of laser surface hardening. Journal of Achievements in Materials and Manufacturing Engineering 2007; 24(1): 17-25.

[9]   Slatter T, Taylor H, Lewis R, King P. The influence of laser hardening on wear in the valve and valve seat contact. Wear 2009; 267(5-8): 797-806.
[http://dx.doi.org/10.1016/j.wear.2009.01.040]

[10]   Pashby IR, Barnes S, Bryden BG. Surface hardening of steel using a high power diode laser. J Mater Process Technol 2003; 139(1-3): 585-8.
[http://dx.doi.org/10.1016/S0924-0136(03)00509-0]

[11]   Jeyaprakash N, Yang CH, Duraiselvam M, Prabu G. Microstructure and tribological evolution during laser alloying WC-12% Co and Cr3C2− 25% NiCr powders on nodular iron surface. Results Phys 2019; 12: 1610-20.
[http://dx.doi.org/10.1016/j.rinp.2019.01.069]

[12]   Selvan JS, Subramanian K, Nath AK. Effect of laser surface hardening on En18 (AISI 5135) steel. J Mater Process Technol 1999; 91(1-3): 29-36.
[http://dx.doi.org/10.1016/S0924-0136(98)00430-0]

[13]   Naeem S, Mehmood T, Wu KM, *et al.* Laser surface hardening of gun metal alloys. Materials (Basel) 2019; 12(16): 2632.
[http://dx.doi.org/10.3390/ma12162632] [PMID: 31430867]

[14]   Moradi M, Ghorbani D, Moghadam MK, Kazazi M, Rouzbahani F, Karazi S. Nd: YAG laser hardening of AISI 410 stainless steel: Microstructural evaluation, mechanical properties, and corrosion behavior. J Alloys Compd 2019; 795: 213-22.
[http://dx.doi.org/10.1016/j.jallcom.2019.05.016]

[15]   Pantelis DI, Bouyiouri E, Kouloumbi N, Vassiliou P, Koutsomichalis A. Wear and corrosion resistance of laser surface hardened structural steel. Surf Coat Tech 2002; 161(2-3): 125-34.
[http://dx.doi.org/10.1016/S0257-8972(02)00495-4]

[16]   Ping X-L, Fu H-G, Wang K-M, *et al.* Effect of laser quenching on microstructure and properties of the surface of track materials. Surf Rev Lett 2018; 25: 1-10.
[http://dx.doi.org/10.1142/S0218625X19500306]

[17]   Wang B, Pan Y, Liu Y, *et al.* Effects of quench-tempering and laser hardening treatment on wear resistance of gray cast iron. J Mater Res Technol 2020; 9(4): 8163-71.
[http://dx.doi.org/10.1016/j.jmrt.2020.05.006]

[18]   Zhang Z, Yang J, Han Y, *et al.* Actively tunable terahertz electromagnetically induced transparency analogue based on vanadium-oxide-assisted metamaterials. Appl Phys, A Mater Sci Process 2020; 126(3): 1-1.
[http://dx.doi.org/10.1007/s00339-020-3374-2]

[19]   Anusha E, Kumar A, Shariff SM. A novel method of laser surface hardening treatment inducing different thermal processing condition for Thin-sectioned 100Cr6 steel. Opt Laser Technol 2020; 125: 106061.
[http://dx.doi.org/10.1016/j.optlastec.2020.106061]

[20]   Oliveira RJ, Siqueira RH, Lima MS. Microstructure and wear behaviour of laser hardened SAE 4130 steels. Int J Surface Sci Eng 2018; 12(2): 161-70.
[http://dx.doi.org/10.1504/IJSURFSE.2018.091231]

[21]   Dinesh Babu P, Buvanashekaran G, Balasubramanian KR. The elevated temperature wear analysis of laser surface–hardened EN25 steel using response surface methodology. Tribol Trans 2015; 58(4): 602-15.
[http://dx.doi.org/10.1080/10402004.2014.998356]

[22]   Zammit A, Abela S, Betts JC, Grech M. Discrete laser spot hardening of austempered ductile iron. Surf Coat Tech 2017; 331: 143-52.
[http://dx.doi.org/10.1016/j.surfcoat.2017.10.054]

[23]   Soriano C, Leunda J, Lambarri J, Navas VG, Sanz C. Effect of laser surface hardening on the microstructure, hardness and residual stresses of austempered ductile iron grades. Appl Surf Sci 2011; 257(16): 7101-6.
[http://dx.doi.org/10.1016/j.apsusc.2011.03.059]

[24]   Shin HJ, Yoo YT, Ahn DG, Im K. Laser surface hardening of S45C medium carbon steel using ND: YAG laser with a continuous wave. J Mater Process Technol 2007; 187: 467-70.
[http://dx.doi.org/10.1016/j.jmatprotec.2006.11.188]

[25]   Babu PD, Buvanashekaran G, Balasubramanian KR. Dry sliding wear of laser hardened low alloy steel at room and elevated temperatures. Proc Inst Mech Eng, Part J J Eng Tribol 2013; 227(10): 1138-49.
[http://dx.doi.org/10.1177/1350650113481530]

[26]   Volpp J, Dewi HS, Fischer A, Niendorf T. Influence of complex geometries on the properties of laser-hardened surfaces. Int J Adv Manuf Technol 2020; 107(9): 4255-60.
[http://dx.doi.org/10.1007/s00170-020-05324-8]

[27]   Molian PA, Baldwin M. Wear Behavior of Laser Surface-Hardened Gray and Ductile Cast Irons: Part 1—Sliding Wear.

[28]   Lesyk DA, Martinez S, Mordyuk BN, *et al.* Combining laser transformation hardening and ultrasonic impact strain hardening for enhanced wear resistance of AISI 1045 steel. Wear 2020; 462: 203494.
[http://dx.doi.org/10.1016/j.wear.2020.203494]

[29]   Hwang JH, Kim DY, Youn JG, Lee YS. Laser surface hardening of gray cast iron used for piston ring. J Mater Eng Perform 2002; 11(3): 294-300.
[http://dx.doi.org/10.1361/105994902770344105]

[30]   Han X, Zhang Z, Pan Y, Barber GC, Yang H, Qiu F. Sliding wear behavior of laser surface hardened austempered ductile iron. J Mater Res Technol 2020; 9(6): 14609-18.
[http://dx.doi.org/10.1016/j.jmrt.2020.10.050]

[31]   Sarkar S, Gopinath M, Chakraborty SS, Syed B, Nath AK. Analysis of temperature and surface hardening of low carbon thin steel sheets using Yb-fiber laser. Surf Coat Tech 2016; 302: 344-58.
[http://dx.doi.org/10.1016/j.surfcoat.2016.06.045]

[32]   Syed B, Shariff SM, Padmanabham G, Lenka S, Bhattacharya B, Kundu S. Influence of laser surface hardened layer on mechanical properties of re-engineered low carbon steel sheet. Mater Sci Eng A 2017; 685: 168-77.
[http://dx.doi.org/10.1016/j.msea.2016.12.124]

[33]   Kapustynskyi O, Višniakov N. Laser treatment for strengthening of thin sheet steel. Adv Mater Sci Eng 2020.
[http://dx.doi.org/10.1155/2020/5963012]

CHAPTER 6

Laser Surface Melting

Muralimohan Cheepu[1,*]

[1] *Super-TIG Welding Co., Limited, Busan- 46722, Republic of Korea*

Abstract: The use of lasers has become a significant source and splendid tool for various surface modifications. Laser surface melting offers extensive promises to accomplish preferred surface properties. The surface is melted by laser irradiation and cooled rapidly by self-quench procedures and is widely utilized as a research tool. The homogeneity of the surface chemical composition is refined and changed by adding other materials in this process. After surface modifications, microstructural changes, corrosion, erosion, and wear properties of different alloys are explained widely in this chapter. The possibilities of repairing the stress corrosion cracking and intergranular corrosion behavior also addressed the effect of heat treatment techniques on stress relief and strengthening of the mechanism of surface treatments.

Keywords: Aging, Aluminum alloys, Annealing, Duplex stainless steel, Ferrite, Hardening, Heat flux, Heat treatment, Inconel, Intergranular corrosion, Intermetallics, Laser, Materials processing, Metallic carbides, Microstructure, Stainless steel, Stress corrosion cracking, Surface modification, Titanium, Wear.

INTRODUCTION

Laser surface melting (LSM) is a flexible and efficient surface treatment process for many industries. It involves heating the required alloys using a laser source. The alloys are melted until the melting point is reached and allowed to solidify quickly. However, it is necessary to avoid the molten surfaces' oxidation during the laser surface melting process [1]. The laser surface melting process is widely used for various materials to enhance their surface properties. Along with the material properties modifications, the importance of laser surface melting has demanded multiple applications. The structural steels of austenitic stainless steel and duplex stainless steels were prone to several cracking and sensitization issues in high-temperature environments [2]. The significant applications of these materials are involved in the chemical, oil and gas, marine, nuclear and construction industries.

---

[*] **Address correspondence to Muralimohan Cheepu:** Super-TIG Welding Co., Limited, Busan-46722, Republic of Korea; Tel: +821027143774; Fax: 05043160402; E-mail: muralicheepu@gmail.com

Jeyaprakash Natarajan and Che-Hua Yang (Eds.)

These materials are highly resistant to corrosion environments and have good mechanical strength in addition to good weldability. However, these alloys are susceptible to sensitization and intergranular corrosion attacks, leading to premature failure of the parts. During welding, due to the heating of the components, sensitization temperature can be reached too high. During the heat treatment processes, it can be affected by heat to cause sensitization [3, 4]. Several methods are implemented to avoid sensitization, such as high-temperature solutionizing and low-temperature treatments using certain rates. The sensitization problem can be solved using laser surface melting by melting the sensitized microstructure at ambient temperature irrespective of the industrial fields, including nuclear industries and complex geometries. Laser surface melting is modest, and economical.

Effective solidification method for the surface of materials prominent to the prolonged solid solution of the alloy structure adds the establishment of various phases and the refined grain structures and the redistribution of secondary phases and the inclusions of effective formation of metastable phases, homogenization and refinement of microstructure, and dissolution/redistribution of precipitates or inclusions. At the same time, the bulk properties can be conserved. Moreover, another significant modification of the alloys is that the remanufacturing of the various components using laser surface melting leads to several benefits, especially in economic aspects. Therefore, the energy consumption and environmental pollution, wastage of precious metals can be reduced [5, 6]. In addition to that, the LSM method can also be applied to various applications where the conventional methods are not feasible to adopt them to repair the damaged parts and the sensitive microstructures and the broken surfaces. It is worth mentioning that the LSM technique has been leading the industry for the last several years to improve the resistance of the large components, which are difficult to install and it also saves the life of equipment from corrosion, erosion, and cavitation problems under the highly compressed air and hot environments [7 - 10]. There are a few more applications that the cracked or precipitated welded joints and their microstructures formed under welding processes [11 - 21] are capable of repairing using the LSM technique. In particular, some of the welded joints have several secondary phases that are responsible for cracks and the failure of the joints [22 - 27], which can be solved by dissolving them using this technique without affecting the properties of the joined structures.

LASER SURFACE MELTING

Laser surface melting involves rapid melting of the surface layer under high-power laser energy sources. The melting and the solidification of metallic materials and other alloys show tremendous potential benefits and advantages.

There are many fundamental investigations underway to understand the relations between laser sources and material behavior. Under such high energy heat sources, metallic materials experience heat flow, fluid flow, and the kinetic molecules of the elements at liquid and solid transactions. The rapid melting and solidification phenomena are different from the normal heating and melting processes. Several heat transfer calculations need to be understood at the solidification of the surface and subsurface layers. The heat flow and heated surface layers at the top and subsurface layer appearance are shown in Fig. (**1**). The layer under the laser beam is exposed to high heat and the molten surface can be seen with the deep depression. The solidified surfaces are slightly affected by adjacent layers, and remelting can occur just before the coating [28]. However, based on the beam diameter, the remelting region has to be decided, as indicated in Fig. (**1**). The heat flow in the thickness direction of the substrate is larger than the sides due to the high energy concentration on the substrate surface, impending the heat source in the downward direction. It is important to understand the heat flow directions and its intensity during laser surface melting, considering critical applications [29 - 31]. The distribution of heat flux absorbed by the substrate is a function of the circular shape's laser spot. It can change due to the reflectivity of the material due to the temperature and time. Fig. (**2**) illustrates the relationship between the uniform heat flux and the Gaussian har flux. The total laser power absorbed by the substrate is Q at the circular shape of the laser spot, which is identical for the Gaussian distribution, and the uniform heat flux is distributed by the given relationship.

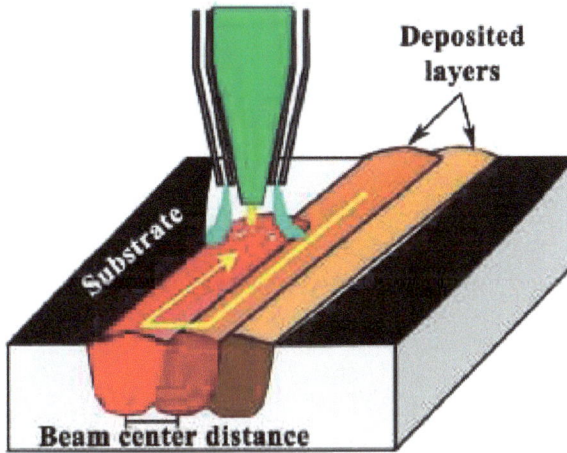

Fig. (1). Schematic view of laser surface melting.

$$q_{uniform} = \frac{q_0}{2.313} \qquad\qquad (1)$$

Where q_0 denotes the absorbed heat flux at the heat zone's laser spot in the Gaussian distribution.

The thermal field solution was subjected to a uniform heat flux for a semi-infinite solid substrate's circular zone. The solution is referenced at the center of the circular region T(0,0), and the relation is shown as follows [32]:

$$\frac{K_s[T(0,0)-T_0]}{q^a} = \frac{2\sqrt{\alpha_s t}}{\sqrt{\pi}a}[1 - \sqrt{\pi}\, ierfc\, \frac{a}{2\sqrt{\alpha_s t}}] \tag{2}$$

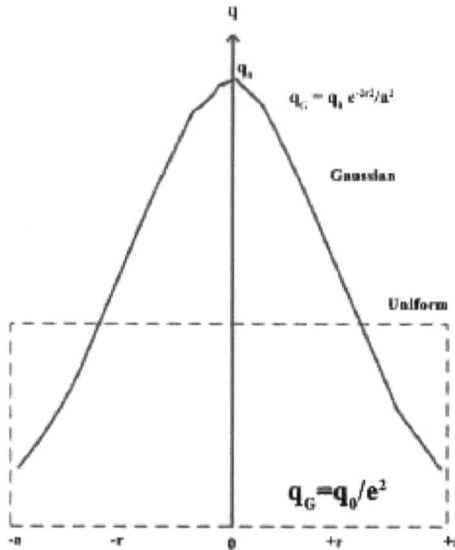

Fig. (2). The relationship between the uniform and Gaussian heat fluxes under the total energy absorbed in the circular laser spot (a is the radius).

The given equation (2) uses the Fig. (2) curve to calculate the vertical axis. The plot exhibits the radius of the laser spot of the circular region 'a' and the required amount of heat flux is needed to reach the center temperature (melting point of the substrate). The given relations were divided into two categories: the small values a/α and the larger Fourier value, which is $\alpha_s t/a^2$.

LASER SURFACE MELTING OF AL-CU ALLOY

The demand for new high-performance materials is growing gradually for critical applications, leading to the development of materials with different alloys. The faster-growing technologies of laser processes for various fields have resulted in the technology for several industrial applications. Aluminum alloys have much importance in regular applications due to their extraordinary properties. However, due to some of their characteristics, they were unable to be used in specific applications.

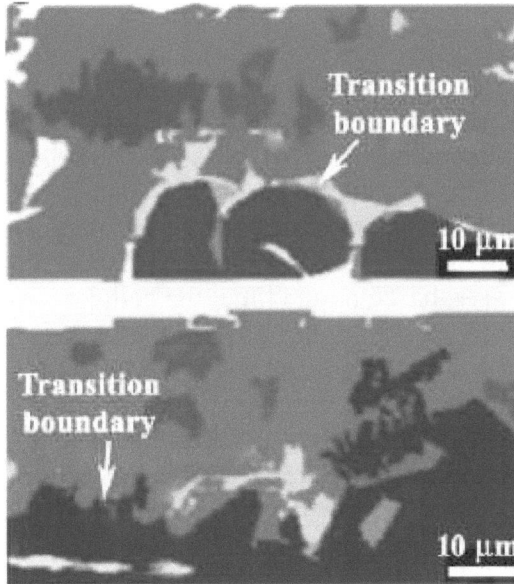

Fig. (3). The morphology shows the molten pool's solidification transitions at different welding speeds (low-speed upper model).

However, it is possible to use them by adding the alloying elements of Cu, Si, and other elements. In recent years, Al alloys are widely used in the parts of the cylinders, mufflers, poser transmission units, crankshaft, and chassis frame that demand lightweight [33]. However, it has less weight and density, which can be used for various lightweight applications. It is necessary to change their properties to enhance the strength and the erosion resistance by using any manufacturing processes. Among them, laser surface melting is the most commonly used process to modify its properties by adding a sufficient amount of copper into the Al alloy [34]. Fig. (3) shows the liquid and solid boundary transition, the resolidified region's microstructure, and the larger grains in the unmelted area. The molten area is larger than the usual molten size due to the long time contact between the beam and the substrates leading to a larger volume. In particular, this alloy's behavior during the laser surface melting showed that the formation of cellular dendritic growth initiated at the bottom of the molten pool and grew towards the pool's top. During this solidification process, dendrites grow directly until their orientation changes. The microstructural evolution along the liquid/solid interface is dependent on the velocity when considering the longitudinal direction of the laser position., and it is defined as the solid/liquid front side (V_s) that combines with the velocity of the beam [35].

$$V_s = V_b \cos \o \tag{3}$$

where ø is the angle between V_s and V_b. The equation explains that V_s' velocity vectors vary from the lower region of the molten pool to the top of the molten pool, whereas V_b can meet the laser track speed [34]. The relation and directions of the V_s and V_b are illustrated in Fig. (4). As shown in Fig. (4), the effect of V_b on the formation of the microstructure is clear, and also it modified the size of the cellular structure. It is also the effect of laser beam speed, and at the lower rates, cellular grains and zone size is more extensive, as shown in Fig. (5) [35]. After growing to a specific size, based on the time, the cellular structure transforms into a dendritic structure under the higher speeds compared to lower speeds. It is exciting that the growth direction changes perpendicular to the V_b at the edge of the solidified bead's tip. It rotates continuously to become parallel to the V_b at the center region of the bead. The cooling rates are much higher in this region, and the solidification rate is also higher along with the molten pool. At this stage, the dendrites lose their growth direction branches and remain cellular. The microstructural formation must have different behavior due to the solidification differences, resulting in the variation of microstructural characteristics. The substrate's average hardness of the unmelted zone is 75 HV, but other zones of the cellular structure have 160 HV hardness.

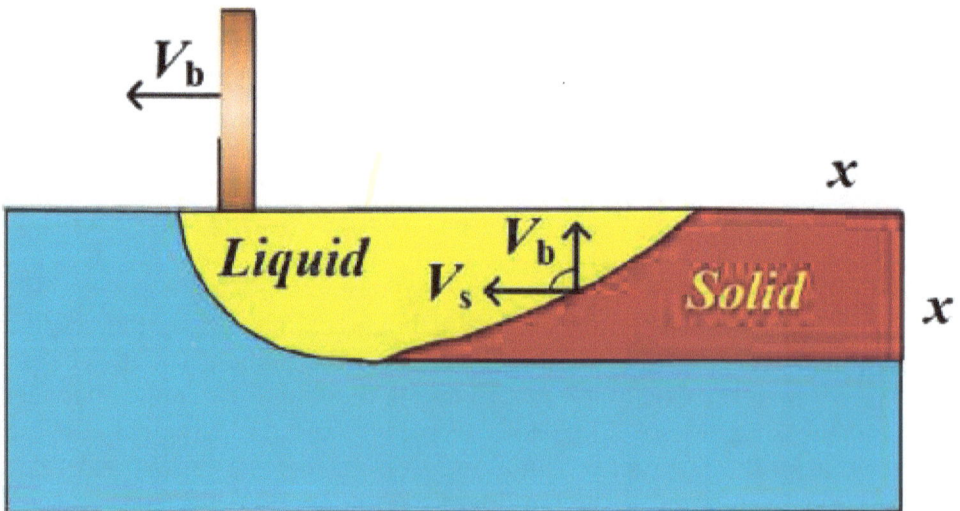

Fig. (4). The vector relationship of the schematic view between the laser beam's speed and solidification speed.

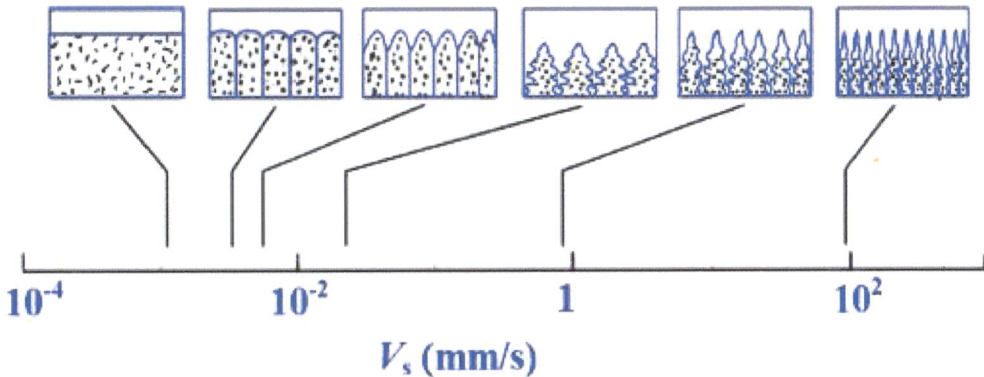

Fig. (5). Schematic microstructural representation for the liquid to solid transition at front velocity.

The highest hardness of 210 HV was obtained for the dendritic structure where the microstructure is refined, and closer arm space is observed between the dendritic structures. The Al-5%Cu alloy's microstructure under the laser surface melting characterized that the three different zones have been formed under the laser track in the longitudinal direction, such as cellular, dendritic structure in the perpendicular direction and parallel direction. If the laser speed changes, the network also changes, and it can be controlled easily by controlling the laser speed. As the laser speed increases, the cellular structure reduces, and the mechanical properties increase [36].

CORROSION AND WEAR BEHAVIOR OF MAGNESIUM ALLOYS AFTER LASER SURFACE MELTING

The laser surface melting technique is also popular to enhance the corrosion, erosion and wear properties of various alloys. The magnesium alloy of AZ91E alloy experienced remarkable improvements in the corrosion and wear properties after LSM surface modifications by applying the SiC and TiC layers on its surface [37, 38]. Moreover, the Mg-2% Zr or 5% Zr alloying substrate's pitting corrosion behavior after LSM performed excellent resistance in the seawater environments. Some of the studies investigated that the Laser Surface Alloying (LSA) of pure commercial magnesium with aluminum, nickel, silicon, copper, and a combination of three elements were found with a substantial improvement in wear resistance [39].

The resistance towards corrosion and wear was high for the variety of Al+Ni layers on the Mg substrate. The magnesium alloy of MEZ type after LSM treatment had good results on pitting corrosion under 3.56 wt.% of NaCl solution. The corrosion rate and the pitting formation were gradually reduced after LSM. The as-received specimen underwent pitting after 30 minutes, but the LSM treated specimen started pitting after the 12 h exposure. Also, the kinetics of the pitting is slower than the as-received substrates. Fig. (**6**) exhibits the immersion time and the area of pits of both the as-received and the LSM substrate. The LSM specimen exhibits a less area of pits due to their area being smaller, and the pit size remains identical after 72 h of the exposure [40].

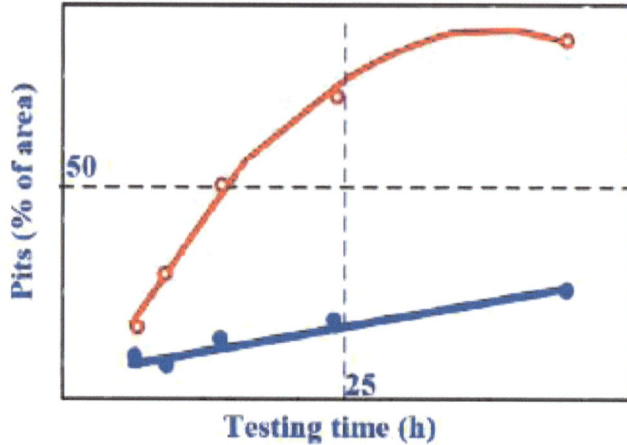

Fig. (6). The relation between the immersion time and the area fraction of the pits of as received (red color) and LSM treated specimen (blue line).

STRESS CORROSION CRACKING REPAIR BY LASER SURFACE MELTING

The stress corrosion cracking (SCC) after the welding and surface treatments leads to significant cracking and substrates failure. The superalloys based on nickel content are extensively used in high environmental plants, such as nuclear and power plants. Cracking/intergranular stress corrosion cracking is a significant threat to the power plants' structural joints [41]. In general, the parts of the nuclear plants were made by cladding using Inconel alloys 182, but they are prone to stress corrosion cracking after decades due to the long-term exposure to the heat and service stresses in the parts. Safety improvement and avoiding damages is necessary to identify these kinds of problems from the parts, and repair is highly needed. During repair or the replacement of these components, efficient methods are used.

The development of practically possible methods needs to be arranged and a process is needed to repair them efficiently. The cracking repairing procedure is essential to avoid the formation of new defects and heat conductivity. The SCC has been restored using LSM in Inconel 182 alloy by the laser beam's specimen surface melting. LSM repaired the preexisting SCC in the parts, but some cracks remain due to the dependability of material, stress, and the environment. Fig. (**7**) shows the two kinds of SCC after being repaired by the LSM; one of them has extended after LSM, and the other one is not developed. The extended crack was induced by the trapped air before LSM in the parts and led to the crack. However, using LSM, the SCC cracking was successfully repaired not only in Inconel alloys but also in many other alloys.

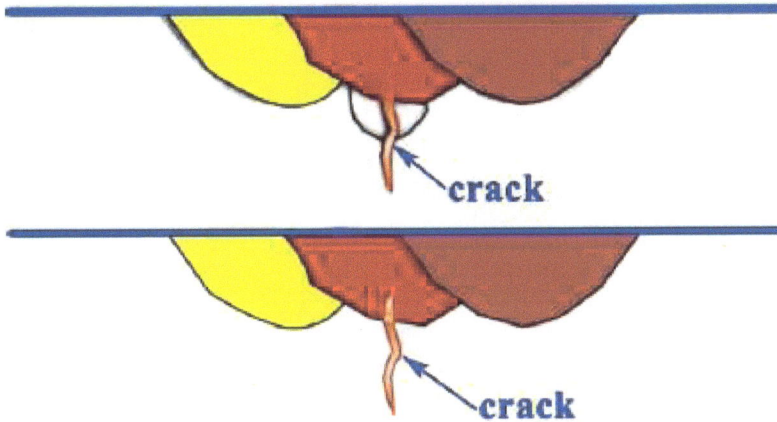

Fig. (7). Stress corrosion cracking behavior after laser surface melting in Inconel 182 alloy.

INTERGRANULAR CORROSION CRACKING REPAIR BY LASER SURFACE MELTING AT DIFFERENT HEAT TREATMENTS

The Fe-Cr-Ni-based alloys of stainless steel and the duplex stainless steels (DSS) are used for various applications. The single-phase austenitic stainless steels (ASS) are prone to intergranular corrosion (IGC) due to the sensitization through the formation of the Cr-rich precipitate of the grain boundaries. When the Cr elemental percentage decreases below 12 percent along the borders where the precipitates were enriched, that zone becomes susceptible to localised corrosion in corrosive conditions [42 - 44]. Simultaneously, the duplex stainless steels are quite different in their microstructure with the presence of two phases, such as 50% ferrite (δ) and 50% austenite (γ) phases. The mechanical strength of the duplex stainless steel is two times higher than that of the ASSs. Also, DSSs have better corrosion resistance than the ASSs due to the higher Cr and Mo consistency. The duplex stainless steels are prone to sensitization at 600 to 950 °C, due to the Cr and Mo rich precipitates and intermetallic sigma phase (σ). It causes the lowering of the mechanical properties and the corrosion resistance of the joints.

In particular, for the alloy of SS31803, the critical pitting temperature and the toughness affected by aging at 850 °C for 10 minutes due to the σ-phase precipitation [45]. There are a few techniques to avoid sensitization by using the heat treatment of solutionizing and the quenching as per the solutionizing range. Several issues arise due to this method with the size and the application of the parts in the nuclear and various industries. Moreover, heating and quenching cause an extensive range of thermal stresses. Therefore, alternative methods are needed, such as laser surface melting, an in-situ process and can repair the sensitization microstructure from the critical regions and seal the cracks for various parts [46]. Fig. (**8**) represents the hardness of the stainless steel specimens treated under different laser surface melting, aged, and annealing conditions. The hardness has been varied between these three heat-treated conditions among the various stainless steels. The hardness at a molten region of the aged samples under LSM showed the same hardness and more than the annealed steels [47 - 51]. The aged ASSs after the LSM treatment necessarily consisted of austenitic along with δ, but the CrC was dissipated. In aged DSSs after LSM, the ferrite (δ) phase is transferred as a significant phase. Moreover, the remaining content of α/γ phase was disturbed, but the phases of σ and γ_2 were wholly removed.

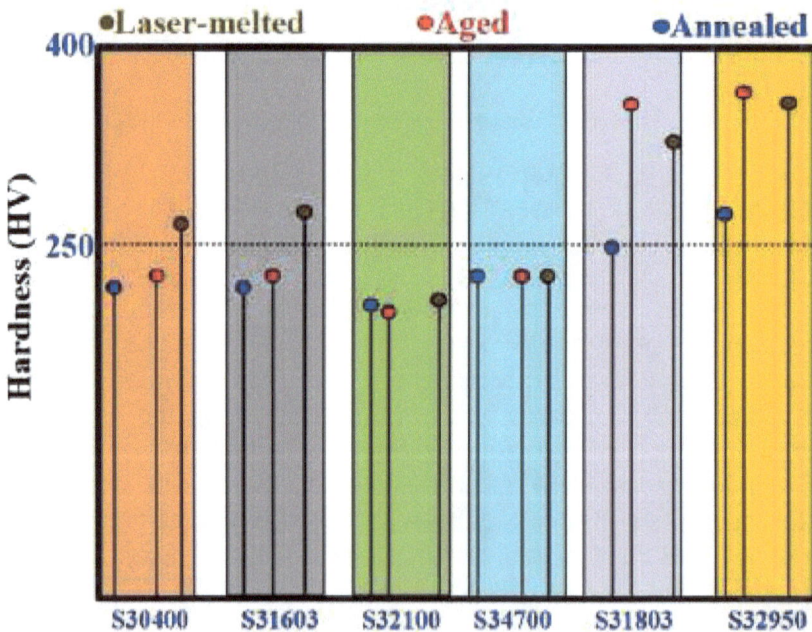

Fig. (8). Microhardness of different stainless-steel grades under other heat treatment conditions and laser surface melting.

CONCLUDING REMARKS

Laser surface melting improves the specific required properties as described above, and there are several other benefits obtained by using this method. In this process, the optimized process parameters are identified for various applications, and the melting process can be applied to any finished components without damaging them. The distortion and the overheating problems are minimal, and significant manufacturing problems are solved using this technique. The wear and corrosion properties were enhanced using laser surface melting process by surface treatment. The hardness and mechanical strength of the substrates were satisfactorily obtained after the melting process. The significant issues of stress corrosion cracking, intergranular corrosion, sensitization, *etc.*, are successfully repaired by sealing them with laser surface melting. The hardness, strength are greatly improved, and the phases are minimized after laser melting in the austenitic and duplex stainless steels. Laser surface melting clearly showed that the surface quality is achievable.

CONSENT FOR PUBLICATION

Not applicable.

CONFLICT OF INTEREST

The author declares no conflict of interest, financial or otherwise.

ACKNOWLEDGEMENTS

The financial support for this work was supported by the Super-TIG welding co., ltd.

REFERENCES

[1]   Paredes RS, Weber FP, Vilar R. Microstructural changes due to laser surface melting of an AISI 304 stainless steel. Mater Res 2001; 4(2): 93-6.
[http://dx.doi.org/10.1590/S1516-14392001000200009]

[2]   Khatak HS, Raj B. Corrosion of Austenitic Stainless Steel: Mechanism, Mitigation and Monitoring. Cambridge, UK: Woodhead Publishing Limited 2002; p. 117.
[http://dx.doi.org/10.1533/9780857094018]

[3]   Dayal RK, Gnanamoorthy JB, Srinivasan G, Esaklul KA. ASM Handbook of Case Histories in Failure Analysis. ASM International. 1993; 2: pp. 253-5.

[4]   Cheepu M, Che WS. Effect of burn-off length on the properties of friction welded dissimilar steel bars. Journal of Welding and Joining 2019; 37(1): 40-5.
[http://dx.doi.org/10.5781/JWJ.2019.37.1.6]

[5]   Liu CY, Yu LY, Tian W, Tang JC. Experiments of laser surface engineering for the green remanufacturing of railway coupler InKey Engineering Materials. Trans Tech Publications Ltd. 2008; 373: pp. 354-7.

[6]   Xu BS. The remanufacturing engineering and automatic surface engineering technology. InKey Engineering Materials. Trans Tech Publications Ltd. 2008; 373: pp. 1-10.

[7]   Conde A, Colaço R, Vilar R, De Damborenea J. Corrosion behaviour of steels after laser surface melting. Mater Des 2000; 21(5): 441-5.
[http://dx.doi.org/10.1016/S0261-3069(00)00037-6]

[8]   Conde A, García I, De Damborenea JJ. Pitting corrosion of 304 stainless steel after laser surface melting in argon and nitrogen atmospheres. Corros Sci 2001; 43(5): 817-28.
        [http://dx.doi.org/10.1016/S0010-938X(00)00114-1]

[9]   Liu Z, Chong PH, Skeldon P, Hilton PA, Spencer JT, Quayle B. Fundamental understanding of the corrosion performance of laser-melted metallic alloys. Surf Coat Tech 2006; 200(18-19): 5514-25.
        [http://dx.doi.org/10.1016/j.surfcoat.2005.07.108]

[10]   Peyre P, Carboni C, Beranger G, *et al.* Influence of laser peening and high power diode laser melting on the pitting corrosion resistance of AISI 316L steel. In International Congress on Applications of Lasers & Electro-Optics 2001 Oct 2001(1): 43-52. Laser Institute of America.

[11]   Krishnaja D, Cheepu M, Venkateswarlu D. A review of research progress on dissimilar laser weld-brazing of automotive applications. In IOP Conference Series: Materials Science and Engineering 2018 Mar 1 330(1): 012073. IOP Publishing. IOP Publishing.
        [http://dx.doi.org/10.1088/1757-899X/330/1/012073]

[12]   Srinivas B, Krishna NM, Cheepu M, Sivaprasad K, Muthupandi V. Studies on post weld heat treatment of dissimilar aluminum alloys by laser beam welding technique. In IOP Conference Series: Materials Science and Engineering 2018 Mar 1 330(1): 012079. IOP Publishing.
        [http://dx.doi.org/10.1088/1757-899X/330/1/012079]

[13]   Anuradha M, Das VC, Venkateswarlu D, Cheepu M. Parameter optimization for laser welding of high strength dissimilar materials. InMaterials Science Forum. Trans Tech Publications Ltd. 2019; 969: pp. 558-64.

[14]   Srinivas B, Cheepu M, Sivaprasad K, Muthupandi V. Effect of gaussian beam on microstructural and mechanical properties of dissimilarlaser welding ofAA5083 and AA6061 alloys. In IOP Conference Series: Materials Science and Engineering 2018 Mar 1 330(1): 012066. IOP Publishing.
        [http://dx.doi.org/10.1088/1757-899X/330/1/012066]

[15]   Cheepu M, Venkateswarlu D, Rao PN, Kumaran SS, Srinivasan N. Effect of process parameters and heat input on weld bead geometry of laser welded titanium Ti-6Al-4V alloy. InMaterials Science Forum 2019 969, pp. 613-618. Trans Tech Publications Ltd.

[16]   Cheepu M, Venkateswarlu D, Rao PN, Kumaran SS, Srinivasan N. Optimization of process parameters using surface response methodology for laser welding of titanium alloy InMaterials Science Forum. Trans Tech Publications Ltd. 2019; 969: pp. 539-45.

[17]   Cheepu M, Srinivas B, Abhishek N, *et al.* Dissimilar joining of stainless steel and 5083 aluminum alloy sheets by gas tungsten arc welding-brazing process. InIOP Conference Series: Materials Science and Engineering 2018 Mar 1, 330(1), p. 012048. IOP Publishing.
        [http://dx.doi.org/10.1088/1757-899X/330/1/012048]

[18]   Cheepu M, Che WS. Friction welding of titanium to stainless steel using Al interlayer. Trans Indian Inst Met 2019; 72(6): 1563-8.
        [http://dx.doi.org/10.1007/s12666-019-01655-7]

[19]   Cheepu M, Venkateswarlu D, Rao PN, Muthupandi V, Sivaprasad K, Che WS. Microstructure characterization of superalloy 718 during dissimilar rotary friction welding. InMaterials Science Forum. Trans Tech Publications Ltd. 2019; Vol. 969: pp. 211-7.

[20]   Anuradha M, Das VC, Susila P, Cheepu M, Venkateswarlu D. Microstructure and Mechanical Properties for the Dissimilar Joining of Inconel 718 Alloy to High-Strength Steel by TIG Welding. Trans Indian Inst Met 2020; 11: 1-5.
        [http://dx.doi.org/10.1007/s12666-020-01925-9]

[21]   Anuradha M, Das VC, Susila P, Cheepu M, Venkateswarlu D. Effect of welding parameters on TIG welding of Inconel 718 to AISI 4140 steel. Trans Indian Inst Met 2020; 11: 1-6.
        [http://dx.doi.org/10.1007/s12666-020-01926-8]

[22]   Muralimohan CH, Muthupandi V, Sivaprasad K. Properties of friction welding titanium-stainless steel

joints with a nickel interlayer. Procedia Materials Science 2014; 5: 1120-9.
[http://dx.doi.org/10.1016/j.mspro.2014.07.406]

[23]   Muralimohan CH, Ashfaq M, Ashiri R, Muthupandi V, Sivaprasad K. Analysis and characterization of the role of Ni interlayer in the friction welding of titanium and 304 austenitic stainless steel. Metall Mater Trans, A Phys Metall Mater Sci 2016; 47(1): 347-59.
[http://dx.doi.org/10.1007/s11661-015-3210-z]

[24]   Muralimohan CH, Haribabu S, Reddy YH, Muthupandi V, Sivaprasad K. Evaluation of microstructures and mechanical properties of dissimilar materials by friction welding. Procedia Materials Science 2014; 5: 1107-13.
[http://dx.doi.org/10.1016/j.mspro.2014.07.404]

[25]   Cheepu M, Muthupandi V, Loganathan S. Friction welding of titanium to 304 stainless steel with electroplated nickel interlayer. InMaterials Science Forum. Trans Tech Publications Ltd. 2012; Vol. 710: pp. 620-5.

[26]   Cheepu M, Ashfaq M, Muthupandi V. A new approach for using interlayer and analysis of the friction welding of titanium to stainless steel. Trans Indian Inst Met 2017; 70(10): 2591-600.
[http://dx.doi.org/10.1007/s12666-017-1114-x]

[27]   Muralimohan CH, Muthupandi V, Sivaprasad K. The influence of aluminium intermediate layer in dissimilar friction welds. International Conference on Engineering Materials and Processes. 350-7.

[28]   Lee KH, Choi SW, Yoon TJ, Kang CY. Microstructure and hardness of surface melting hardened zone of mold steel, SM45C using Yb: YAG disk laser. Journal of Welding and Joining 2016; 34(1): 75-81.
[http://dx.doi.org/10.5781/JWJ.2016.34.1.75]

[29]   Choi D, Shiu J. Weld Shape Analysis using Central Composite Design in the Laser Welding of Aluminum Alloys. Journal of Welding and Joining 2020; 38(5): 502-7.
[http://dx.doi.org/10.5781/JWJ.2020.38.5.10]

[30]   Yun T-J, Oh W-B, Lee B-R, *et al.* A Study on Optimization of Fillet in Laser Welding Process for 9% Ni Steel Using Gradient-Based Optimization Algorithm. Journal of Welding and Joining 2020; 8(5): 485-92.
[http://dx.doi.org/10.5781/JWJ.2020.38.5.8]

[31]   Kim J-Y, Lee D-M. LAM-DED Process for Repair and Maintenance of Cast Iron Components using Metallic Powder Alloys. Journal of Welding and Joining 2020; 38(4): 349-58.
[http://dx.doi.org/10.5781/JWJ.2020.38.4.3]

[32]   Mehrabian R, Kou S, Hsu SC, Munitz A. Laser surface melting and subsequent solidification. InAIP Conference Proceedings 1979 Apr 15 50(1), pp. 129-148. American Institute of Physics.
[http://dx.doi.org/10.1063/1.31653]

[33]   Mietz J. Materials science and engineering–An introduction. Von WD Callister, Jr., 3. Auflage, XX, 811 S., zahlreiche Abb. und Tab., John Wiley & Sons. Inc. New York, 1994, paper back 37.95,handback 79.50, ISBN 0 471 30568 5.

[34]   Pinto MA, Cheung N, Ierardi MC, Garcia A. Microstructural and hardness investigation of an aluminum–copper alloy processed by laser surface melting. Mater Charact 2003; 50(2-3): 249-53.
[http://dx.doi.org/10.1016/S1044-5803(03)00091-3]

[35]   Zimmermann M, Carrard M, Kurz W. Rapid solidification of Al-Cu eutectic alloy by laser remelting. Acta Metall 1989; 37(12): 3305-13.
[http://dx.doi.org/10.1016/0001-6160(89)90203-4]

[36]   Kühn GW. Kurz, DJ Fisher, Fundamentals of Solidification. Trans Tech Publications 1986.

[37]   Hiraga H, Inoue T, Kamado S, Kojima Y. Improving the wear resistance of a magnesium alloy by laser melt injection. Mater Trans 2001; 42(7): 1322-5.
[http://dx.doi.org/10.2320/matertrans.42.1322]

[38] Kim C, Kim J, Lee H, *et al.* Effect of Laser Power Feedback Control on Mechanical Properties of Stainless Steel Part Built by Direct Energy Deposition. Journal of Welding and Joining 2020; 38(2): 197-202.

[39] Galun R, Weisheit A, Mordike BL. Improving the surface properties of magnesium by laser alloying. Corros Rev 1998; 16(1-2): 53-74.
[http://dx.doi.org/10.1515/CORRREV.1998.16.1-2.53]

[40] Majumdar JD, Galun R, Mordike BL, Manna I. Effect of laser surface melting on corrosion and wear resistance of a commercial magnesium alloy. Mater Sci Eng A 2003; 361(1-2): 119-29.
[http://dx.doi.org/10.1016/S0921-5093(03)00519-7]

[41] Hoang PH, Gangadharan A, Ramalingam SC. Primary water stress corrosion cracking inspection ranking scheme for alloy 600 components. Nucl Eng Des 1998; 181(1-3): 209-19.
[http://dx.doi.org/10.1016/S0029-5493(97)00346-4]

[42] Pillai SR, Khatak HS. Corrosion of austenitic stainless steel in liquid sodium. InCorrosion of Austenitic Stainless Steels. Woodhead Publishing. 2002; pp. 241-64.
[http://dx.doi.org/10.1533/9780857094018.265]

[43] Cheepu M, Susila P. Interface microstructure characteristics of friction-welded joint of titanium to stainless steel with interlayer. Trans Indian Inst Met 2020; 7: 1-5.
[http://dx.doi.org/10.1007/s12666-020-01895-y]

[44] Cheepu M, Che WS. Characterization of microstructure and interface reactions in friction welded bimetallic joints of titanium to 304 stainless steel using nickel interlayer. Trans Indian Inst Met 2019; 72(6): 1597-601.
[http://dx.doi.org/10.1007/s12666-019-01612-4]

[45] Deng B, Wang Z, Jiang Y, Sun T, Xu J, Li J. Effect of thermal cycles on the corrosion and mechanical properties of UNS S31803 duplex stainless steel. Corros Sci 2009; 51(12): 2969-75.
[http://dx.doi.org/10.1016/j.corsci.2009.08.015]

[46] Kaul R, Parvathavarthini N, Ganesh P, *et al.* A novel pre-weld laser surface treatment for enhanced inter-granular corrosion resistance of austenitic stainless steel weldment. Weld J 2009; 88: 233s-42s.

[47] Kwok CT, Lo KH, Chan WK, Cheng FT, Man HC. Effect of laser surface melting on intergranular corrosion behaviour of aged austenitic and duplex stainless steels. Corros Sci 2011; 53(4): 1581-91.
[http://dx.doi.org/10.1016/j.corsci.2011.01.048]

[48] Jeyaprakash N, Yang CH, Duraiselvam M, Sivasankaran S. Comparative study of laser melting and pre-placed Ni–20% Cr alloying over nodular iron surface. Arch Civ Mech Eng 2020; 20(1): 1-2.
[http://dx.doi.org/10.1007/s43452-020-00030-4]

[49] Jeyaprakash N, Yang CH, Duraiselvam M, Prabu G, Tseng SP, Kumar DR. Investigation of high temperature wear performance on laser processed nodular iron using optimization technique. Results Phys 2019; 15: 102585.
[http://dx.doi.org/10.1016/j.rinp.2019.102585]

[50] Cho WI, Na SJ. Impact of Wavelengths of CO_2, Disk, and Green Lasers on Fusion Zone Shape in Laser Welding of Steel. Journal of Welding and Joining of Welding and Joining 2020; 38(3): 235-40.
[http://dx.doi.org/10.5781/JWJ.2020.38.3.1]

[51] Lee TH, Oh JH, Kam D-H. High Speed Photography for Arc Welding Phenomenon Using 808 nm Diode Laser Illumination and Optical Filter. Journal of Welding and Joining 2020; 38(5): 429-34.
[http://dx.doi.org/10.5781/JWJ.2020.38.5.1]

CHAPTER 7

Laser Surface Alloying

Venkateswarlu Devuri[1,*] and **Muralimohan Cheepu**[2]

[1] *Department of Mechanical Engineering, Marri Laxman Reddy Institute of Technology and Management, Hyderabad, India*

[2] *Super-TIG Welding Co., Limited, Busan - 46722, Republic of Korea*

Abstract: The material processing procedure, which utilizes a high-power density laser, was made to focus on a metal coating to melt the thin layer of the substrate is called Laser Surface Alloying (LSA). It is one of the efficient processes to fabricate material for proper microstructure and non-equilibrium solidification. These fabricated materials have shown good corrosion and wear resistance than the base material of several alloys. Compared to traditional methods like welding, thermal spraying, *etc.*, LSA had the advantage of the constrained heat-affected zone (HAZ), high density, and good mechanical properties. Surface modification using laser alloying of various materials, processing techniques, properties of the alloying surfaces, and combination of multiple alloys and elements to the laser surface modification are described in detail, along with the favorable conditions for adhesive wear.

Keywords: Alloying surface, Aluminum, Ceramics, Cracks, Dilution, Hardness, Heat affected zone, Inconel alloys, Interface, Laser surface alloying, Mechanical properties, Metallic glass, Microhardness, Powders, Porosity, Process techniques, Titanium, Solidification, Stainless steel, Wear.

INTRODUCTION

The present global applications used in industries involve resources with unique surface assets that showcase extraordinary hardness and confrontation to wear. The alloys that inherit these properties are not so economical, leading to an increase in parts' price. By manufacturing the materials with LSA makes production economical, and these also inherit properties similar to alloys made traditionally, and they can also withstand high surface stresses [1]. By following the mentioned process, distinctive combinations can be obtained, like the high

---

* **Correspondence author Venkateswarlu Devuri:** Department of Mechanical Engineering, Marri Laxman Reddy Institute of Technology and Management, Hyderabad, India; Tel: +918885337628; Fax: 914029556182; E-mail: devuri.venky@gmail.com

Jeyaprakash Natarajan and Che-Hua Yang (Eds.)

hardness of the surface accompanied by the bulk's higher impact strength. Laser Surface Alloying (LSA) contains either a depositing alloying element as a powder in dry form and the suspension onto the base substrate surface by brushing with hand gas jet, *etc.*, (or) by creating a layer of alloying elements at the substrate surface and melting the added layer partly with the underlying substrate by laser. Intermixing of the substrate material and alloying material takes place in a thin surface layer. Thus, the alloyed layer forms at the substrate's surface. The surface properties of the material could be affected by a combination of distinct alloying materials. There will be a creation of intermetallic products through a reaction between the base and alloying materials chemically. The creation of these compounds in the solid phase consists of 2 or more metallic elements in distinct proportions that characterize material hardness. The parameters involved in this process are mentioned below:

1. Laser power
2. Beam diameter
3. Laser scanning speed
4. Powder feed rate

These parameters must be kept correctly to acquire planned or designed properties. All the parameters and their role is shown in Fig. (1). Any uncontrolled parameter may lead to undesired properties of output. This technology creates an immensely dense and fracture-free product and exhibits an outstanding bond with the base material. The effects generated by LSA yield high resistance to wear at different temperatures. This also allows a vast range of coating products for applications in industries.

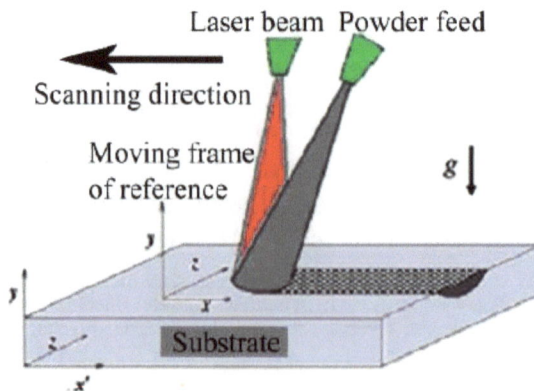

Fig. (1). Schematic view of the laser process with the scanning direction and the coordinates of the substrate.

IMPORTANCE OF LASER SURFACE ALLOYING (LSA)

Laser surface alloying is mainly known for its wear resistance, which can be used in certain areas like tooling. This is also accessed in applications where corrosion resistance of base material is required; this process increases resistance to oxidation of base material. It is done in materials that are compatible with one other.

The main advantage of LSA is that the dilution of base material into alloying material is minimal or less compared to other processes. Heat input can be made significantly lower by giving proper laser beam parameters. The lower dilution of alloying material into base material allows alloying material to regain its original properties. The Heat Affected Zone (HAZ) is also less due to minimal heat input; this also keeps base material properties with minimal change. The lower HAZ causes a faster cooling rate making the product with high hardness also with good resistance to wear.

The main benefits of the LSA include:

1. Minimal heat input.
2. Minimal change in mechanical properties as there is a small change in the microstructure of base material.
3. Quality and properties can be changed based on input parameters.

This shows that laser coating has significant benefits over the welding and cladding process.

MECHANISM OF LASER ALLOYING

The metal surfacing process that utilizes the source of the high-energy laser beam is called laser alloying. First of all, the alloying powder has been spread on the base sheet or material; after this, the laser is projected on the powder to melt it and form an intermetallic bond between the base and alloying material. The process method and the powder feeding mechanism are indicated in Fig. (**2**).

IMPACT OF LASER PROCESSING PARAMETERS

There are various factors or parameters influencing or managing the laser process. Each factor does have its distinct purpose, which could affect the outcome of the process, mainly on the properties of the material. The main factors which influence the properties are laser size, rate of powder flow, laser power, and laser speed. These factors help in achieving a thin cladding layer and promote good strength. They also influence temperature distribution and good quality micro-structure.

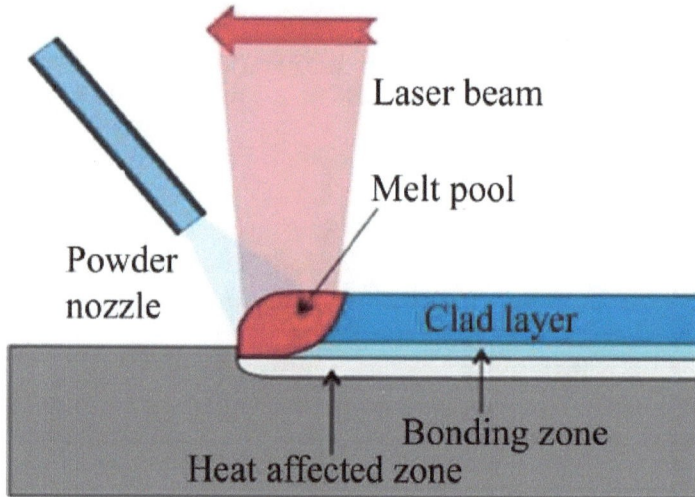

Fig. (2). Laser surface alloying process and substrate modifications with the deposited multilayers.

For example, a study conducted by Vaziri et al. proved that Al alloyed with Ni showed an improved hardness up to 10 times the value of base Al material. Laser power reduction had shown a reduction in clad layer thickness.

Laser Scanning Speed

The speed referring to the speed of laser travelling along the workpiece is laser scanning speed. Higher or faster speeds can result in a constant reduction in the quantity of cladding material onto the workpiece, making a very thin layer of alloying material.

Laser Power

The crucial factor influencing the whole process is the laser power, as it is accountable for energy density being transferred on the powder substrate. This affects the microstructure of the fabricated product.

Alloying Powder Flow Rate

The rate of flow of powder determines the thickness of the clad layer and also the rate of dilution required. This majorly tries to affect change in microstructure, surface finish, cracks, and porosity.

LASER ALLOYING OF DISTINCT MATERIALS

Aluminum

The alloys manufactured using aluminum are the widely used nonferrous materials in industrial applications as they have a higher specific strength, even though its poor hardness and lower wear-resistant have reduced its use in specific industrial applications. To improve the mechanical properties of the surface, various technology has been introduced, applied; it was found that the modifications of the surface by adding alloys through laser process have attracted many industrialists recently. The resistance of this material towards cavitation erosion is poorer because of its poor hardness in the surface and low strength [2]. When the alloy of aluminum is subjected to larger impact stresses, it forms liquid cavities or localized failure due to fatigue. To overcome these failures, we need to clad a good alloy with aluminum which is known for its cavitation resistance [3]. The perfect match for this is NiCrSiB alloy, which has higher hardness and excellent wear properties.

A remarkable improvement in surface hardness has been noticed in the laser cladded layer on the surface of the base material. The given figure represents the profile of hardness value *vs.* height or distance from the surface [4]. A value of around 900 Hv has been noticed in the alloyed layer at the top of the surface. By controlling the laser parameters, as mentioned before, the hardness value may be altered, which also influences the manufacturing cost. Also, higher processing speeds may lead to an increase in the cooling rates of cladded material. This also demonstrates the weighing effect of the rate of cooling and speeds upon microstructure and surface hardness of cladded layers.

The hardness has been seen to be decreased with an increase in depth of the cladded layer from the top layer, as shown in Fig. (**3**). The next figure represents the laser alloyed NiCrSiB onto aluminum anodic polarization behavior. It shows the potential to resist corrosion that has been improved drastically compared to the regular base material. As compared to Al6061, the potential for the protection of laser cladded material has shifted to direction notably to y700 mV from y900 mV. This is due to the presence of nickel, the intermetallic bond formed between nickel-aluminum and chromium-aluminum. The presence of heterogeneous surfaces has shown a lack of passivation, which includes distinct phases that could result in local galvanic cells. A similar result has been noticed when aluminum is alloyed with ceramic Silicon-Nitrate, as in Fig. (**4**).

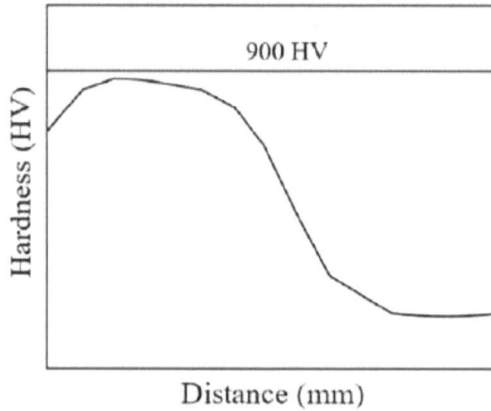

Fig. (3). Microhardness of the laser alloying surfaces.

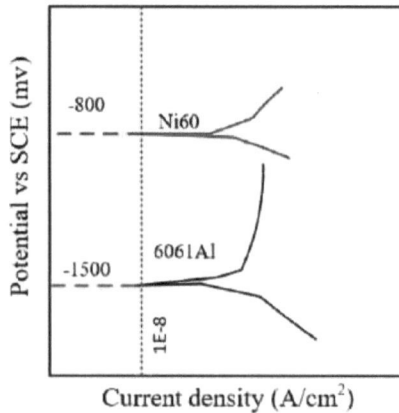

Fig. (4). The curves for the aluminum 6061 alloy and Nickel 60 show the anodic polarization trend and current density.

Titanium

It is an accessible metal in the crust of the earth, making it the 4th most available metal and it is denoted by Ti. The most known titanium alloy is Ti-6Al-4V, popularly used in industrial applications like aerospace, airframe, chemical and power generation industries. It has very fine corrosive and mechanical properties [5]. Some of the distinctive properties of Ti and alloys of Ti are lower density, amazing weight to strength ratio, good specific strength even at high temperatures [6]. It also has excellent resistance to corrosion, low elastic modulus, and good flexibility. The titanium is alloyed as it has low shear and tensile strength. It also has a high coefficient of friction. The laser surface alloyed titanium with TiN-S--B-Ni showed a huge improvement in the properties of the material compared to pure Ti [7]. Fig. (**5**) represents the hardness of laser cladded alloy of Ti compared to pure Titanium.

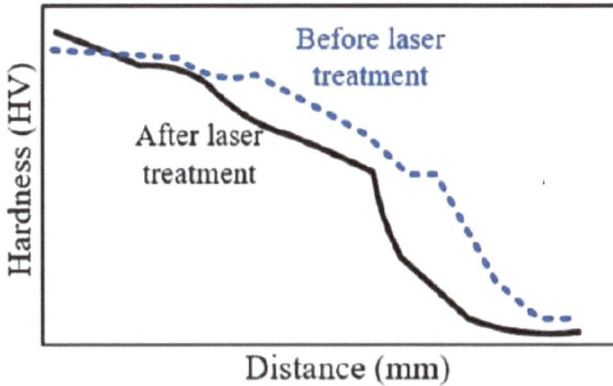

Fig. (5). Hardness profiles of the titanium alloy before and after laser alloying.

The hardness of the laser cladded Ti is observed to be 1600 HV, which is higher than the pure Ti. This significantly proved that the usage of laser alloyed materials in industries can be more efficient than pure Ti. Other experiments have revealed a difference in the coefficient of friction between pure and laser-clad Ti. The laser alloyed Ti has a lower coefficient of friction than pure Ti, which plays an important role in industrial applications [8]. Therefore, this also shows that laser alloyed materials play a crucial role in producing efficient products.

Stainless Steel

There are several grades of stainless steel, such as austenitic, ferritic and martensitic, *etc*. The prominent properties of stainless steel are strength and resistance to corrosion. Among the various stainless steel, austenitic grade steels became more popular to be used for various applications. In particular, the austenitic steels are categorized based on their chemical compositions and physical properties. The most universally used alloy type is 316L type steel (Cr-Ni-Mo). The presence of Cr and Mo along with Ni achieved the highest ductility and enhanced the allowable limit for forming techniques. Based on the applications, the parts were prepared using several manufacturing methods. Some of the reliable and standard parts in the automotive industries were chosen to prepare the special alloys using powder metallurgy techniques. It can reduce the wastage of material and manpower. The deceptive disadvantage of steels manufactured by powder metallurgy route is that they have lower mechanical strength owing to essential porosity [9]. Consequently, the surface properties of the steels are also critical when they are exposed to corrosion media. Hence, it is required to treat them using surface modification techniques to expand surface hardness. Among the others, the laser surface technique is a remarkable method for powder metallurgy steels. The benefit of the laser surface method is the interaction of a small, focused area by the laser beam. Laser surface alloying is a method with rigid particles to increase wear resistance. The hardness, surface properties and wear resistance properties increase gradually by applying the laser surface alloying method. The studies on 316L stainless steel with Inconel 625 powder deposition results in the increase of mechanical and corrosion properties [10 - 19].

Duplex Stainless Steel

One of the many types of steel is duplex stainless steel, which is commonly known as DSS. It has an ASTM A240 grade, which is similar or equivalent to S32205, known as Duplex 2205. It consists of austenite and ferrite [20, 21]. Even though having high corrosion resistance, it lags due to its lower hardness and lower resistance to wear. The major alloying processes used for DSS are laser alloying and thermal spraying. When DSS is laser cladded with SiC, it shows a huge improvement in hardness up to 1135 HV as compared to 220 HV of the substrate. It also shows an improvement in resistance to wear and corrosion.

Superalloys

A superalloy is one of the utmost prevalent types of alloys that are used for modern manufacturing processes. The excellent corrosion resistance and hardness of the alloys made them different from the other alloys. But the alloying elements during addition show that the Al or steel might decline and not lever heat well; Inconel can withstand high temperatures [22, 23]. Inconel alloys are ideal for transforming their phases in extreme environments. These are widely used for various high-temperature applications such as chemical, food processing and nuclear engineering. Inconel 617 is one of the superalloys, which can be used for the gas turbine engine. In particular, it is most popular for chemical processing applications owing to its high resistance to wear and corrosion [24]. Inconel 625 is another grade of superalloy. It has been employed in cryogenic range temperature due to its excellent oxidation-resistant. It has wide applications, such as furnace hardware and combustion liners [25 - 27]. For laser surface heat treatments, Inconel 718 alloy is mostly used because it has shown significant properties towards the thermal stresses. Based on some experiments conducted and with the help of published journals, we can clearly conclude that using Inconel 718 alloy has improved the surface hardness of the material 2~3 times more than the substrate material using laser surfacing. Laser surface hardening also can change the alloying elements of dissimilar materials based on the applications as dissimilar materials welding [28 - 45].

Metallic Glass

Metallic glasses are amorphous, which refers to metallic alloys that have a short order of atomic structure. In the practical world, they are formed by conserving the metal quench to make disordered configurations essential to conformist crystalline metals. Metallic glasses show better strength, splendid corrosion and wear owing to the nonconventional atomic arrangement in them. The non-equilibrium methods used to make these alloys confine their geometry with the tens of millimeters and due to this geometry, their applications are limited to small-sized components and 2-d coatings.

NiCr Powder

Fig. (**6**) shows the NiCr powder size and its morphology with the presence of uniform particle size. The cross-sectional view of the substrate after laser surface alloying is presented in Fig. (**7**). The microstructure of the alloying morphology showed that the formation of uniform distribution of the particles with the proper solidification. The solidification structure of dendrite formation is seen in Fig. (**8**). Furthermore, wear tests on the alloyed surface revealed that it has outstanding wear resistance, as represented in Fig. (**9**) [46 - 50].

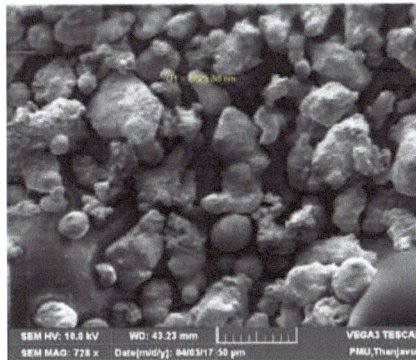

Fig. (6). The microstructure of the NiCr powder.

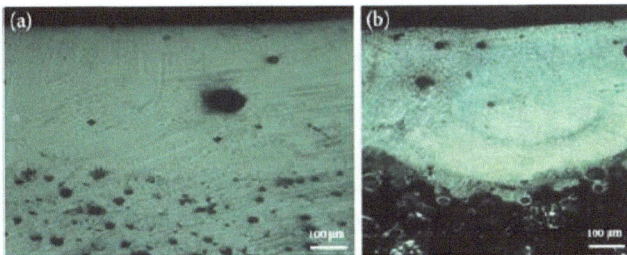

Fig. (7). The cross-sectional view of the NiCr coated substrates by (a) conventional coating technique (b) LSA technique.

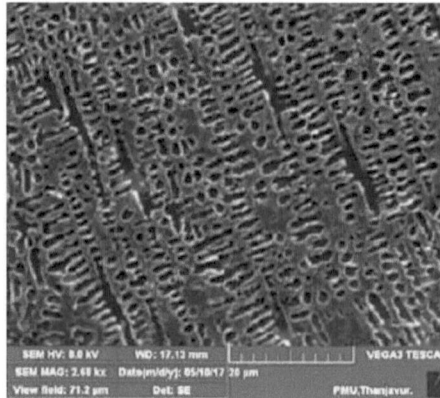

Fig. (8). SEM image shows the solidification morphology of the NiCr alloying substrate by laser surface alloying.

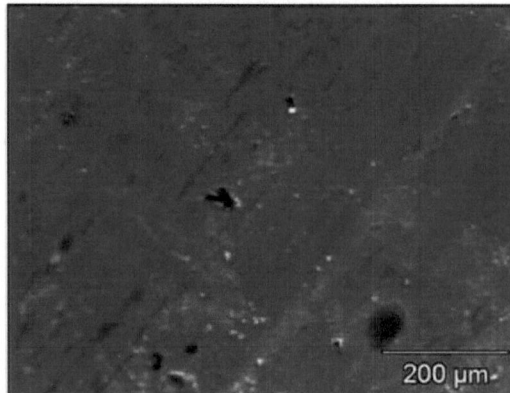

Fig. (9). SEM image shows the NiCr worn-out surface after the laser surface alloying.

PROPERTIES OF LASER SURFACE ALLOYING

Laser surfacing involves heating of base metal to a temperature where the poured material (deposited material) can form an efficient bonding. Generally, by this process, the surface hardness will increase much more times compared with the parent metal. In general, we use alloying to maximize the performance of the materials than their performance in original conditions.

An alloyed material shows better performance than the original/pure material. Generally, alloying is performed to enhance strength, hardness, wear resistance and to reduce the costs for further required properties. Alloying of each material has its significance. Based on applications and required properties, we select the alloying element for the parent base metal. For example, the final required properties will entirely depend on the alloyed materials. Titanium alloys are used for high toughness applications. Nickel-based alloys show better properties in high-temperature applications. Steels are used for producing better surface hardness, wear and corrosion resistance, but the amount of carbon in the steel decides the degree of enhancement in properties. The most desirable steel to use with the laser process is Alloy Steel. Alloying elements are especially molybdenum, manganese, boron, and chromium. These steels are used for heat treatment of range 3 mm depth without any concern of back tempering.

ADVANTAGES AND APPLICATIONS

Nowadays, the usage of lasers is rapidly increasing. They started their applications only for unique purpose substitutions for the replacement of many traditional methods.

They serve the following advantages:

- Enhanced mechanical properties.
- Thermal and heat resistance.
- Aesthetic look.
- Enhanced machinability and fabricability, *i.e.*, high productivity and control.
- The formed part is free of crack and porosity, *i.e.*, high quality of the alloyed layer.
- Due to the high cooling rate, a fine microstructure is obtained.
- To prepare/produce any machine part, the surface geometry is not important. Any kind of complex geometry can be made easily.
- This increases the lifetime of the tool and the machined components.
- Nowadays, lasers are used in 90% of the mechanical industries to have a fast and better production rate.
- Laser surface-coated polymers are mostly used in biomedical applications. They are used for various treatments to reduce the cost and time for the surgeries.
- Laser surface alloyed parts are mostly used in aircraft and turbine jets due to their better experimental results. In aircraft and turbine jets, the worn-out components can easily modify or replace using surface modification techniques.
- They have contributed to many other applications like electronic materials, nanotechnology, nuclear reactors/energy.

CONCLUDING REMARKS

Laser surface alloying is a suitable method for enhancing the material's surface strength of various metals by adding hard surfaces. There are several methods developed to perform this operation and improve the industrial part's surface properties. Most current processes can build the parts and their surfaces using layer-wise techniques with exact geometries. Almost every metal, ceramics, and glasses can be manufactured easily using this technique. It is the most economical method and is easy to apply for the various parts, with the dimensions and complexity of the final cut being practically limitless. The dilution of base material into alloying material is minimal or less compared to other processes. The main factors that influence the properties are laser size, rate of powder flow, laser power and laser speed. The substrate characteristics of the laser preserved layer achieved the greatest improvements, and alloyed layers obtained superior corrosion resistance.

CONSENT FOR PUBLICATION

Not applicable.

CONFLICT OF INTEREST

The author declares no conflict of interest, financial or otherwise.

ACKNOWLEDGEMENTS

The financial support for this work was supported by the Marri Laxman Reddy Institute of Technology and Management, India.

REFERENCES

[1]   Kusiński J, Kac S, Kopia A, *et al.* Laser modification of the materials surface layer-a review paper. Bull Pol Acad Sci Tech Sci 2012; 60(4): 711-28.
[http://dx.doi.org/10.2478/v10175-012-0083-9]

[2]   Viscusi A, Leitão C, Rodrigues DM, Scherillo F, Squillace A, Carrino L. Laser beam welded joints of dissimilar heat treatable aluminium alloys. J Mater Process Technol 2016; 236: 48-55.
[http://dx.doi.org/10.1016/j.jmatprotec.2016.05.006]

[3]   Tomida S, Nakata K, Ushio M. Trends in welding research Proceedings of 5th International Conference. Georgia, USA. 1998; p. 478.

[4]   Tomida S, Nakata K. Fe–Al composite layers on aluminum alloy formed by laser surface alloying with iron powder. Surf Coat Tech 2003; 174: 559-63.
[http://dx.doi.org/10.1016/S0257-8972(03)00698-4]

[5]   Adeleke SA. TIG Melted Surface Modified Titanium Alloy for automotive cylinder liner application. International Journal of Automotive Engineering and Technologies 2015; 4(3): 130-8.
[http://dx.doi.org/10.18245/ijaet.57450]

[6]   Bello KA, Maleque MA, Adebisi AA, Dube A. Preparation and characterisation of TIG-alloyed hybrid composite coatings for high-temperature tribological applications. Transactions of the IMF. 211-.
[http://dx.doi.org/10.1080/00202967.2016.1182727]

[7]   Maleque MA, Ghazal BA, Mohammad YA, Hayyan M, Ahmed AS. Wear behaviour of TiC coated AISI 4340 steel produced by TIG surface melting. InMaterials Science Forum. Trans Tech Publications Ltd. 2015; 819: pp. 76-80.

[8]   Cai Z, Jin G, Cui X, *et al.* Synthesis and microstructure characterization of Ni-Cr-Co-Ti-V-Al high entropy alloy coating on Ti-6Al-4V substrate by laser surface alloying. Mater Charact 2016; 120: 229-33.

[http://dx.doi.org/10.1016/j.matchar.2016.09.011]

[9]   Yan J, Gao M, Zeng X. Study on microstructure and mechanical properties of 304 stainless steel joints by TIG, laser and laser-TIG hybrid welding. Opt Lasers Eng 2010; 48(4): 512-7.
[http://dx.doi.org/10.1016/j.optlaseng.2009.08.009]

[10]   Khalfallah IY, Rahoma MN, Abboud JH, Benyounis KY. Microstructure and corrosion behavior of austenitic stainless steel treated with laser. Opt Laser Technol 2011; 43(4): 806-13.
[http://dx.doi.org/10.1016/j.optlastec.2010.11.006]

[11]   Maleque A, Karim R. Tribological behavior of dual and triple particle size SiC reinforced Al-MMCs: a comparative study. Ind Lubr Tribol 2008.
[http://dx.doi.org/10.1108/00368790810881533]

[12]   Mridha S, Dyuti S. Effects of Processing Parameters on Microstructures and Properties of TIG Melted Surface Layer of Steel. InAdvanced Materials Research. Trans Tech Publications Ltd. 2011; 264: pp. 1421-6.

[13]   Mridha S, Baker TN. Incorporation of 3 μm SiCp into Titanium surfaces using a 2.8 kW laser beam of 186 and 373 MJ m− 2 energy densities in a nitrogen environment. J Mater Process Technol 2007; 185(1-3): 38-45.
[http://dx.doi.org/10.1016/j.jmatprotec.2006.03.110]

[14]   Dyuti S, Mridha S, Shaha SK. Surface modification of mild steel using tungsten inert gas torch surface cladding. Am J Appl Sci 2010; 7(6): 815.
[http://dx.doi.org/10.3844/ajassp.2010.815.822]

[15]   Maleque MA, Bello KA. M Idriss AN, Mirdha S. Processing of TiC-CNT hybrid composite coating on low alloy steel using TIG torch technique. InApplied Mechanics and Materials. Trans Tech Publications Ltd. 2013; 378: pp. 259-64.

[16]   Mridha S, Md Idriss AN, Maleque MA, Yaacob II, Baker TN. Melting of multipass surface tracks in steel incorporating titanium carbide powders. Mater Sci Technol 2015; 31(11): 1362-9.
[http://dx.doi.org/10.1179/1743284714Y.0000000712]

[17]   Emamian A, Corbin SF, Khajepour A. Effect of laser cladding process parameters on clad quality and in-situ formed microstructure of Fe–TiC composite coatings. Surf Coat Tech 2010; 205(7): 2007-15.
[http://dx.doi.org/10.1016/j.surfcoat.2010.08.087]

[18]   Brytan Z, Bonek M, Dobrzański LA, Pakieła W. Surface layer properties of sintered ferritic stainless steel remelted and alloyed with FeNi and Ni by HPDL laser. InAdvanced Materials Research. Trans Tech Publications Ltd. 2011; Vol. 291: pp. 1425-8.

[19]   Buytoz S. Microstructural properties of SiC based hardfacing on low alloy steel. Surf Coat Tech 2006; 200(12-13): 3734-42.
[http://dx.doi.org/10.1016/j.surfcoat.2005.01.106]

[20]   Charles J. Duplex Stainless Steels-a Review after DSS '07 held in Grado. Steel Res Int 2008; 79(6): 455-65.
[http://dx.doi.org/10.1002/srin.200806153]

[21]   Paijan LH, Berhan MN, Adenan MS, Yusof NF, Haruman E. Structural development of expanded austenite on duplex stainless steel by low temperature thermochemical nitriding process. InAdvanced Materials Research. Trans Tech Publications Ltd. 2012; Vol. 576: pp. 260-3.

[22]   Ram GJ, Reddy AV, Rao KP, Reddy GM. Microstructure and mechanical properties of Inconel 718 electron beam welds. Mater Sci Technol 2005; 21(10): 1132-8.
[http://dx.doi.org/10.1179/174328405X62260]

[23]   Shi B, Attia H, Vargas R, Tavakoli S. Numerical and experimental investigation of laser-assisted machining of Inconel 718. Mach Sci Technol 2008; 12(4): 498-513.
[http://dx.doi.org/10.1080/10910340802523314]

[24] Liu L, Hirose A, Kobayashi KF. Laser surface annealing technique of aged Inconel 718 by laser-beam irradiation. InFirst International Symposium on High-Power Laser Macroprocessing 2003 Mar 3, 4831, pp. 241-246. International Society for Optics and Photonics.
[http://dx.doi.org/10.1117/12.497895]

[25] Luo X, Yoshihara S, Shinozaki K, Kuroki H, Shirai M. Theoretical analysis of grain boundary liquation in heat affected zone of Inconel 718 alloy. Study of laser weldability of Ni-base superalloys (3rd Report). Welding international 2000; 14(11): 865-73.

[26] Xiao M, Poon C, Wanjara P, Jahazi M, Fawaz Z, Krimbalis P. Optimization of Nd: YAG-laser welding process for inconel 718 alloy. InMaterials science forum. Trans Tech Publications Ltd. 2007; 546: pp. 1305-8.

[27] Antonsson T, Fredriksson H. The effect of cooling rate on the solidification of INCONEL 718. Metall Mater Trans, B, Process Metall Mater Proc Sci 2005; 36(1): 85-96.
[http://dx.doi.org/10.1007/s11663-005-0009-0]

[28] Krishnaja D, Cheepu M, Venkateswarlu D. A review of research progress on dissimilar laser weld-brazing of automotive applications. InIOP Conference Series: Materials Science and Engineering 2018 Mar 1, 330(1), p. 012073). IOP Publishing.
[http://dx.doi.org/10.1088/1757-899X/330/1/012073]

[29] Srinivas B, Krishna NM, Cheepu M, Sivaprasad K, Muthupandi V. Studies on post weld heat treatment of dissimilar aluminum alloys by laser beam welding technique. InIOP Conference Series: Materials Science and Engineering 2018 Mar 1, 330(1), p. 012079. IOP Publishing.
[http://dx.doi.org/10.1088/1757-899X/330/1/012079]

[30] Anuradha M, Das VC, Venkateswarlu D, Cheepu M. Parameter optimization for laser welding of high strength dissimilar materials. InMaterials Science Forum. Trans Tech Publications Ltd. 2019; Vol. 969: pp. 558-64.

[31] Srinivas B, Cheepu M, Sivaprasad K, Muthupandi V. Effect of gaussian beam on microstructural and mechanical properties of dissimilarlaser welding ofAA5083 and AA6061 alloys. InIOP Conference Series: Materials Science and Engineering 2018 Mar 1, 330(1), p. 012066. IOP Publishing.
[http://dx.doi.org/10.1088/1757-899X/330/1/012066]

[32] Cheepu M, Venkateswarlu D, Rao PN, Kumaran SS, Srinivasan N. Effect of process parameters and heat input on weld bead geometry of laser welded titanium Ti-6Al-4V alloy. InMaterials Science Forum 2019 (Vol. 969, pp. 613-618). Trans Tech Publications Ltd.

[33] Cheepu M, Venkateswarlu D, Rao PN, Kumaran SS, Srinivasan N. Optimization of process parameters using surface response methodology for laser welding of titanium alloy. InMaterials Science Forum. Trans Tech Publications Ltd. 2019; Vol. 969: pp. 539-45.

[34] Cheepu M, Srinivas B, Abhishek N, *et al.* Dissimilar joining of stainless steel and 5083 aluminum alloy sheets by gas tungsten arc welding-brazing process. InIOP Conference Series: Materials Science and Engineering 2018 Mar 1, 330(1), p. 012048. IOP Publishing.
[http://dx.doi.org/10.1088/1757-899X/330/1/012048]

[35] Cheepu M, Che WS. Friction welding of titanium to stainless steel using Al interlayer. Trans Indian Inst Met 2019; 72(6): 1563-8.
[http://dx.doi.org/10.1007/s12666-019-01655-7]

[36] Cheepu M, Venkateswarlu D, Rao PN, Muthupandi V, Sivaprasad K, Che WS. Microstructure characterization of superalloy 718 during dissimilar rotary friction welding. InMaterials Science Forum. Trans Tech Publications Ltd. 2019; 969: pp. 211-7.

[37] Anuradha M, Das VC, Susila P, Cheepu M, Venkateswarlu D. Microstructure and Mechanical Properties for the Dissimilar Joining of Inconel 718 Alloy to High-Strength Steel by TIG Welding. Trans Indian Inst Met 2020; 1-5.
[http://dx.doi.org/10.1007/s12666-020-01925-9]

[38]   Cheepu M, Che WS. Effect of burn-off length on the properties of friction welded dissimilar steel bars. Journal of Welding and Joining 2019; 37(1): 40-5.
[http://dx.doi.org/10.5781/JWJ.2019.37.1.6]

[39]   Anuradha M, Das VC, Susila P, Cheepu M, Venkateswarlu D. Effect of welding parameters on TIG welding of Inconel 718 to AISI 4140 steel. Trans Indian Inst Met 2020; 11: 1-6.
[http://dx.doi.org/10.1007/s12666-020-01926-8]

[40]   Muralimohan CH, Muthupandi V, Sivaprasad K. Properties of friction welding titanium-stainless steel joints with a nickel interlayer. Procedia Materials Science 2014; 5: 1120-9.
[http://dx.doi.org/10.1016/j.mspro.2014.07.406]

[41]   Muralimohan CH, Ashfaq M, Ashiri R, Muthupandi V, Sivaprasad K. Analysis and characterization of the role of Ni interlayer in the friction welding of titanium and 304 austenitic stainless steel. Metall Mater Trans, A Phys Metall Mater Sci 2016; 47(1): 347-59.
[http://dx.doi.org/10.1007/s11661-015-3210-z]

[42]   Muralimohan CH, Haribabu S, Reddy YH, Muthupandi V, Sivaprasad K. Evaluation of microstructures and mechanical properties of dissimilar materials by friction welding. Procedia Materials Science 2014; 5: 1107-13.
[http://dx.doi.org/10.1016/j.mspro.2014.07.404]

[43]   Cheepu M, Muthupandi V, Loganathan S. Friction welding of titanium to 304 stainless steel with electroplated nickel interlayer. InMaterials Science Forum. Trans Tech Publications Ltd. 2012; 710: pp. 620-5.

[44]   Cheepu M, Ashfaq M, Muthupandi V. A new approach for using interlayer and analysis of the friction welding of titanium to stainless steel. Trans Indian Inst Met 2017; 70(10): 2591-600.
[http://dx.doi.org/10.1007/s12666-017-1114-x]

[45]   Muralimohan CH, Muthupandi V, Sivaprasad K. The influence of aluminium intermediate layer in dissimilar friction welds International Conference on Engineering Materials and Processes. 350-7.

[46]   Jeyaprakash N, Yang CH, Duraiselvam M, Sivasankaran S. Comparative study of laser melting and pre-placed Ni–20% Cr alloying over nodular iron surface. Arch Civ Mech Eng 2020; 20(1): 1-2.
[http://dx.doi.org/10.1007/s43452-020-00030-4]

[47]   Jeyaprakash N, Duraiselvam M, Aditya SV. Numerical modeling of WC-12% Co laser alloyed cast iron in high temperature sliding wear condition using response surface methodology. Surf Rev Lett 2018; 25(07): 1950009.
[http://dx.doi.org/10.1142/S0218625X19500094]

[48]   Jeyaprakash N, Duraiselvam M, Raju R. Modelling of Cr3C2-25% NiCr laser alloyed cast iron in high temperature sliding wear condition using response surface methodology. Arch Metall Mater 2018; 63: 63.

[49]   Jeyaprakash N, Yang CH, Duraiselvam M, Prabu G. Microstructure and tribological evolution during laser alloying WC-12% Co and Cr3C2– 25% NiCr powders on nodular iron surface. Results Phys 2019; 12: 1610-20.
[http://dx.doi.org/10.1016/j.rinp.2019.01.069]

[50]   Jeyaprakash N, Yang CH, Tseng SP. Optimization of tribological parameters over WC-12% Co laser alloyed pearlitic ductile iron using Taguchi based Grey relational analysis. J Cent South Univ 2020; 27(3): 736-51.
[http://dx.doi.org/10.1007/s11771-020-4327-9]

CHAPTER 8

Laser Cladding: Process Parameter, Characterization and Defect Analysis – Review and Future Trends

R. Dinesh Kumar[1], **Nandhini Ravi**[2, *] and **Varthini Rajagopal**[3]

[1] *Department of Mechanical Engineering, Indian Institute of Technology, Guwahati-781039, Assam, India*

[2] *Department of Metallurgical and Materials Engineering, National Institute of Technology, Tiruchirappalli-620015, Tamil Nadu, India*

[3] *Department of Mechanical Engineering, Government College of Engineering Srirangam-620012, Tamil Nadu, India*

Abstract: Laser surface treatments for surface modification, parts renewal, and manufacturing of complex/near-net-shaped components are the recent research hotspots. Laser cladding is one such bulk deposition coating technique, where an amalgamation of materials with desirable properties is melted using a laser energy source and deposited over a moving substrate. Once the deposited material cools and solidifies, a clad layer is formed on the substrate resulting in strong metallurgical bonding inducing elevated heat resistance coating and tribological properties of the meshing surfaces. This chapter reviews the influences of various laser cladding process parameters, such as laser power, scan speed, beam diameter, powder feeding methods/rate, beam focal position on the cladding geometry, dilution rate, layer thickness, aspect ratio, microstructure, and tribological properties. Then, the defects observed in laser cladding techniques are reviewed, along with the causes and the remedies reported in the literature. Finally, the tribological applications of laser cladding in traditional and novel materials are also reported.

Keywords: Applications, Cooling rates, CO_2 laser, Defects, Energy Density, Fiber laser, Focus Height, Functionally graded material layers, Laser Cladding, Laser Power, Microcracks, Microhardness, Nd:YAG laser, Porosity, Post Heat Treatment, Powder feed rate, Process Parameters, Remedies, Scan Speed, Wear Resistance.

* **Address correspondence to Nandhini Ravi:** Department of Metallurgical and Materials Engineering, National Institute of Technology, Tiruchirappali-620015, Tamil Nadu, India; Tel: +91-9790779692; Email: nandhiniravi20@gmail.com;

INTRODUCTION

In laser cladding, the deposition of the filler material on the substrate surface is achieved by melting the filler material using a high-power laser source. This method is used for obtaining a new surface and for repairing damaged or worn-out surfaces. Minimal melting of the base material produces 100% metallurgical joint and is hence used in various industries, including aerospace, shipbuilding, automobile, transport, oil and gas, power engineering, *etc* [1]. Laser cladding is achieved through two major techniques, namely a remelting technique that uses a preplaced material over the surface for cladding (Fig. **1**) and a fusion cladding technique that uses external powder or wire feed to clad over the material surface (Fig. **2**).

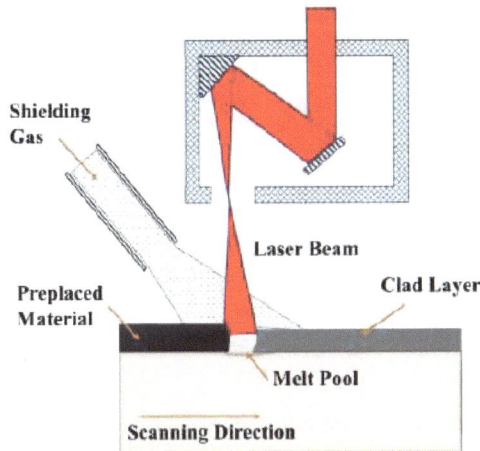

Fig. (1). Remelting laser cladding technique.

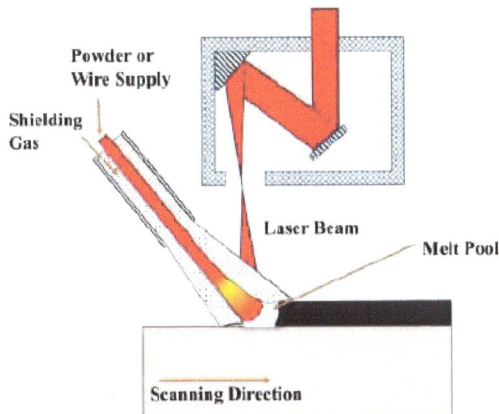

Fig. (2). Fusion laser cladding technique.

Laser cladding and surface alloying are surface modification methods employed to produce thin coating/layer with enriched surface properties or to repair surface defects by developing highly resistant gradient coatings/layers on the substrate. These techniques are found to be appropriate to process a wide range of materials due to the high energy density and cooling rates. Laser cladding uses different types of laser including CO_2 laser of wavelength 10.6 microns, Nd:YAG laser of wavelength 1.06 microns, fibre laser of wavelength 1.07 microns, and direct diode laser of wavelength 0.8-1 micron [2]. The advancement of high power lasers in recent times enhanced control and delivery mechanisms, thereby attracting extensive research in laser surface treatment [3]. Colloidal graphite powder and pure nickel powder were used in laser cladding to obtain a TiC reinforced Ni-based coating on Ti-6Al-4V.

The improvement in microhardness of the coating was explained due to the solid solution strengthening mechanism, which had occurred by the diffusion of the substrate atoms, leading to substantial *in situ* growth of TiC in the shape of the polyhedron and petalite [4]. *In-situ* formation of a laser-cladded layer on Ti-6Al-4V titanium alloy was performed to repair underwater. Both the layers were obtained underwater and in-air reported columnar β-grains, which consisted of the lamellar α phase and acicular α′ phase. The water-cooling effect in the underwater environment led to fine grain size and thin α lamellae and enhanced the amount of α′ phase. A cleavage fracture mechanism was observed in the underwater layer with river pattern features [5]. Several coating techniques, such as physical vapor deposition, chemical vapor deposition, thermal spraying, plasma spraying, electrochemical process, laser cladding, can be adapted to produce these self-lubricating wear-resistant coatings. The cutting-edge improvements in producing self-lubricating coatings by laser cladding have been reviewed and studied [6]. There are also different types of feed methods in the cladding technique, as given in Fig. (**3**).

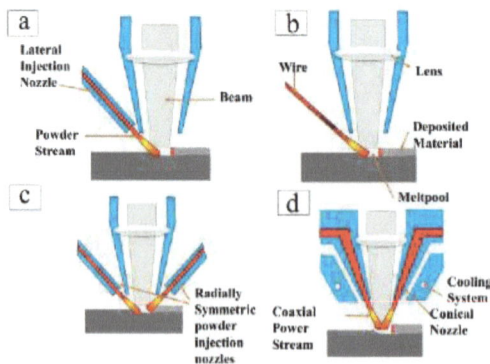

Fig. (3). Different types of feed methods in laser cladding.

A new technique of a one-step method of laser cladding, using paste bound hard facing alloys, is presented. This method allows the location of hard facing alloy precisely with respect to the laser beam and is, therefore, able to control the melting conditions [7]. In AlCoCrFeNiSi$_x$ high-entropy alloy coating was developed on AISI 304 stainless steel, where dislocation strengthening was found to contribute quantitatively for the enhancement of microhardness [8, 9]. Substantial temperature gradient and fast circular convection occur in the molten pool while the molten pool attains a steady state in a very short duration. The molten pool exhibited columnar growth at the bottom, equiaxed growth at the top, and columnar to equiaxed transition at the centre [10]. An increase in laser scan rate or decrease in laser power decreased the depth of the molten substrate [11]. Several methods of welding methodologies have been discussed to provide a Nitinol surface clad on a substrate. The applicability of laser cladding in gas turbine engines has also been explored to repair parts, such as vanes, stators, seals, and rotors [12]. The hard facings obtained by laser cladding at low dilution were found to exhibit improved wear resistance, compared to that of the same hard facings obtained by plasma spraying [13]. The coating of Ni over 42CrMo reported higher microhardness than that of the substrate. Ni45 + 10%Mo composite coating exhibited wear resistance 1.734 times of Ni45 coating and 2.367 times of 42CrMo steel [13]. The thermal expansion coefficient of the clad coating was also found to be lesser than that of the substrate at temperatures in the range of 30 °C to 500 °C [14]. Inert gas or vacuum atmosphere should be strictly maintained to prevent the oxidation of magnesium alloys. Factors influencing the wettability of the coating with magnesium substrate need to be studied with prime importance [15]. The laser cladding process has been found to be advantageous compared to other conventional methods by the appropriate selection of process parameters, laser optics, power, and type of laser [1, 16]. However, the coating possesses a wider passivation region due to refined grains and the high content of passivation elements [17]. Further studies can be done on this temperature monitoring method by investigating the influence of workpiece size and the influence of substrate preheating on the temperature gradient of the laser cladding process [18, 19].

IMPACTS OF PROCESS PARAMETERS

Laser cladding comprises a variety of process parameters, such as traversing speed, laser power, and powder feed rate, which were varied during laser cladding of Ti-6Al-4V clad powder on Ti-6Al-4V substrate. Their microstructures were obtained, which ranged from a thin martensite structure to a thick dendritic martensite structure. Laser traversing speed, when decreased, produced a refined Widmanstatten microstructure at the heat-affected zone. Coarse martensite microstructure was reported for high laser power [20]. High tensile residual stresses are induced during laser cladding, which can be detrimental to the substrate material. Hence, it becomes essential to study the formation mechanism and assess the distribution of residual stresses in the clad layer. Functionally graded material layers were prepared on Ti6Al4V sheet substrate by laser cladding for this purpose. The residual stresses induced in the clad layers were found to be less than the yield stress of the clad material. This confirms that fractures may not be induced by the laser cladding process. The residual stress distributions were also found to be influenced by the material properties. Two aspects of material properties were revealed; a lower coefficient of thermal expansion relieved residual stresses by decreasing shrinkage during cooling, while a higher Young's modulus increased residual stresses by magnifying stress equivalent to strain [21]. By the introduction of a buffer layer which is found to be having a composition common to both the substrate and the clad layer, a steep hardness gradient from the clad layer to the substrate could be overcome. One such work has been reported, which involved depositing a Stellite 6 clad layer with a buffer layer of Inconel 625 on SS316 substrate.

Clad layers with and without buffer layer were obtained at varying heat inputs to investigate the influence of the buffer layer. Clad height was found to increase at lower heat inputs when clad with a buffer layer. The fine-grained microstructure was obtained at the interface when claded with buffer layer while coarse-grained microstructure when claded directly. The maximum hardness of 578 HV was obtained in the clad layer with a buffer layer at low heat input because of the fine-grained microstructure and high carbide content. This, in turn, led to an improvement in the wear rate. It was thus concluded that laser cladding of Stellite 6 with Inconel 625 buffer layer at low heat input (135 J/mm) presented outstanding microstructural and mechanical properties [22]. The analysis of thermal barrier coatings of zirconia and alumina obtained by laser cladding was performed on Udimet 700 superalloy and AISI 4140 low alloy steel substrates to obtain the high-temperature performance efficiency of these materials. CO_2 gas lasers with 1.2- and 5-kW power levels were utilized for cladding. Thin zirconia clads of 5-15 μm could be attained, which were found to be having substantially high microhardness in the range of 800-1700 $HV_{0.2}$, very fine microstructure, free of cracks, and tremendous adhesion bonding. Thicker clads were found possible to be produced, but cracks occurred and were delaminated from the substrate. The dominant process variable was found to be the power density, which determined the formation of a thick, cracked clad or a thin, crack-free clad. Further progress in obtaining thick flawless clads would be possible by obtaining the relationship between the power density and the mechanism of clad pool formation [16]. An *in-situ* synthesis reaction combined with laser cladding on a mild steel substrate was investigated. Tungsten, carbon, and nickel powder particles were placed on a slot of the mild steel substrate, after which laser cladding was performed with argon gas shielding. Such an *in-situ* developed clad showed improved mechanical performance than the conventionally developed clad with tungsten carbide powder particles. In an *in-situ* developed clad, the nucleation and growth of tungsten carbide particles occur in the molten pool, creating an excellent interfacial bonding with the matrix.

The homogenous distribution of Tungsten Carbide particles was confirmed in the *in-situ* developed clad. At higher loads during wear testing, tungsten carbide particles in the conventional clad layer were easily pulled off due to the weaker interfacial bonding than the *in-situ* developed clad [18]. Co42 + B_4C combination powders were coated on Ti-6Al-4V alloy using laser cladding. The clad layer obtained was found to comprise of γ-Co/Ni, TiC, TiB_2, TiB, NiTi, CoTi, $CoTi_2$, and Cr_7C_3 phases. Refinement of TiB particles and TiC dendrites were found to occur in the microstructure while the laser-specific energy was decreased from 12.7 kJ cm^{-2} to 4.9 kJ cm^{-2}. Microhardness and wear resistance were further improved by this refinement [23]. Superalloy Ni35 powder was clad on a Q235 steel substrate using a fibre laser. The effect of defocussing (30 mm and 50 mm) was investigated on the mechanical properties of the clad layer. With an increase in defocus, it is reported that a finer microstructure was obtained, increasing the hardness and corrosion resistance of the clad layer. Upon heat-treating the clad workpiece up to 600 °C, the grains were found to further refine, leading to the enhancement of the hardness of the clad layer and confirming the stability of the clad specimen at high-temperature environments. The clad layer obtained at high defocusing was found to perform superiorly [24]. The effect of post mechanical treatment on the performance of laser clad coating has also been attempted. In one such work, ultrasonic impact treatment was performed on the post-laser clad surface of $Al_{0.5}CoCrFeMnNi$ high entropy material. A skinny impacted layer with plastic deformation was obtained upon performing ultrasonic impact treatment once. While treatment was done twice, a still finer microstructure layer was formed. Upon increasing the number of ultrasonic impact treatments, impact tracks and micro-cracks began to form because of the excessive plastic deformation undergone. It was reported that the phase structure of the clad layer, which originally consisted of the FCC phase, was unaltered after ultrasonic impact treatment. Improvement in surface roughness, surface hardness, and corrosion resistance was confirmed for one ultrasonic impact treatment [25]. The effect of the presence of graphite in the laser clad layer was investigated by cladding Fe313 powder on the Grey Cast Iron HT250 substrate material. Thermo-mechanical modelling of this laser cladding process was performed. Graphite present in cladding region dissolved completely while graphite present in bonding region dissolved partially.

As per the model, tensile stresses were introduced in the substrate during cladding, which increased gradually in the cooling stage. This stress concentration accumulates at the graphite tips, thereby being prone to the initiation of microcracks. The dimensions of graphite played a major role in the tensile stress concentration at the graphite tip. It is reported that the tensile stresses increased with an increase in graphite length and radius [26]. The process of laser cladding integrated with the cooling stage was attempted by fabrication of the IN718 clad layer on an IN718 substrate immersed in liquid nitrogen using a diode laser. It was observed that a very rapid cooling rate occurred in laser cladding because of liquid nitrogen. In this integrated laser cladding cooled by liquid nitrogen, the Laves phase was found to refine more than that of the conventional laser clad layer. Nb segregation in the Laves phase upon solidification was also decreased during laser cladding cooled in liquid nitrogen. When this clad was subjected to heat treatment, additional Nb was found to distribute in austenite and precipitate into γ'-Ni_3Nb, which, in turn, improved the mechanical performance of the clad surface [27]. The beam interaction time, also termed the dwell time, plays a major influence on the solidification and microstructure of the clad layer. Such an analysis was carried out by cladding Stellite 6H on a 12Cr-Ni turbine blade substrate. A short dwell time allowed the clad zone to exhibit a higher hardness due to the fine microstructure, while a long dwell time made the clad zone have a lower hardness due to the coarse microstructure. A short dwell time was also found to considerably decrease the dilution of the substrate material into the clad layer. This allows the clad to retain its strength after long exposure in turbine blade environments [28]. The influence of process parameters, such as scan speed, laser power, and powder feeding rate, was investigated on the clad dimensions of Fe-based alloy YCF104 powder on YCF104 substrate. From the orthogonal experimental design adapted for this investigation, it is proven that laser power had a major influence on the depth and width of heat-affected zone and scan speed had a major influence on the height of the clad layer. It is also revealed that the powder feeding rate had a negligible influence on the obtained clad dimensions [29].

A fuzzy modelling approach and optimization were applied to laser cladding of Stellite 6 powder on the EN8 steel substrate. The process parameters, such as laser power, scan speed, focus height, and powder feed, were taken into consideration. The corrosion behaviour of the clad layer obtained at various process parameter combinations was recorded as electron microscopy images which were fed into the fuzzy logic model. From the fuzzy modelling approach and the further optimization method of Taguchi, the optimum process parameter combination for the minimum corrosion index output was determined as 1800 W laser power, 8 mm/min scan speed, and 0.1 g/min powder feed [30]. The effect of laser path overlap was studied on surface roughness and microhardness of laser clad Ni-C-

-Si-B alloy on 304 stainless steel substrate using a diode laser. A hardness of 11% higher than that of the base powder clad material was obtained at the laser scan speed of 20 mm/s and for laser path overlap of less than 60%. With an increase in laser path overlap, a much flatter clad layer was produced, which was evident with the decrease in surface roughness values. At a laser path overlap of 70%, the microhardness of the clad layer was found to be nearly equivalent to that of the base powder clad material [31]. The effect of laser power on the properties of NiCrBSi clad prepared on 42CrMo steel substrate using fibre laser was investigated. The depth of penetration of the clad into the substrate surface was observed to increase with an increase in laser power.

The phases present in the clad layer were identified to be γ-Ni, (Fe,Ni), $M_{23}C_6$, M_7C_3 and CrB, where M=Fe, Cr. M_7C_3 and $M_{23}C_6$ are hard in nature. The microstructure at the interface was columnar, while at the middle of the clad was cellular in structure. Excellent bonding between the clad coating and substrate was attained at different power values. The hardness of the clad coating was found to decrease with an increase in laser power, while the wear resistance of the clad layer was increased first and then decreased. At laser power of 2000 W, the wear resistance of the clad coating was obtained to be the maximum [32]. The influence of laser-specific energy on the performance of Ni-Ti-Cr powder clad coated by laser cladding on Q235 mild steel was studied. The clad coating consisted mainly of phases NiTi, Cr and Cr_2Ti. *in-situ* synthesis of NiTi happened during the laser cladding, which was proved to enhance the crack resistance of the clad coating. The clad coating was not bonded with the substrate at the lowest laser-specific energy. When laser-specific energy was maintained at 50 J/mm^2, the clad coating recorded a higher microhardness, which was more than that of the mild steel substrate. At lower and higher laser-specific energy, the clad coating could not attain optimal wear resistance [33].

OPTIMIZATION TECHNIQUES

Optimization techniques are used to predict the optimal parameter and validate with the experimental data. Recent researchers have used several optimization techniques. Response surface methodology was applied in the cladding of WC powder on Inconel-718 substrate to determine the role of process parameters in achieving a defect-free clad layer. With a high pulse width, high laser power, and low laser scan speed, the presence of cracks in the clay layer was decreased. The depth of penetration of clad material in the substrate increased, and hence, dilution increased with an increase in laser power. The optimum combination was obtained as 173 W laser power, 12.4 ms pulse width, and 2.9 mm/s laser scan speed. Pull-off strength of 94.24 MPa and microhardness of 1211 HV0.3 were achieved at the optimum parameter combination using a laser path overlap of 50%

[34]. A fuzzy modelling approach and optimization were applied to laser cladding of Stellite 6 powder on the EN8 steel substrate. The process parameters, such as laser power, scan speed, focus height, and powder feed were taken into consideration. The corrosion behaviour of the clad layer obtained at various process parameter combinations was recorded as electron microscopy images, which were fed into the fuzzy logic model. From the fuzzy modelling approach and the further optimization method of Taguchi, the optimum process parameter combination for the minimum corrosion index output was determined as 1800 W laser power, 8 mm/min scan speed, and 0.1 g/min powder feed [30]. A single-track layer of 316 stainless steel powders was clad on 316 stainless steel substrate material with which the thermal imager was also integrated to determine the influence of process parameters on the temperature distribution during cladding. Finite element modelling was also performed, and the comparison of the simulation and the experimental results were obtained. Upon increasing laser scan speed, clad height, width, and melt pool temperature were reported to decrease. Upon increasing laser power, clad width and melt pool temperature were observed to increase. Laser power was found to have minimal influence on clad height. This model was found to be useful in understanding the temperature distribution during the laser cladding at different periods. This model can further be applied to different materials for temperature distribution analysis [35].

The optimization of process parameters by a particle swarm algorithm was performed on laser cladding of pre-placed $CoCr_{1.5}FeNiNb_{0.5}$ powders on 32CrNi2MoVA substrate material using a fibre laser. Process parameters chosen were laser power, laser scan rate, and defocus, while the output variables were residual stress and dilution. Laser scan speed was reported to be the dominant parameter for residual stress, while defocus played a significant linear effect on dilution. The clad coating was found to consist of a face-centered cubic structure and the Laves phase. The clad coating exhibited a hardness of 600 Hv [36]. Taguchi optimization approach was adapted in laser 0cladding Mg-4.2Zn-1.2-e-0.7Zr filler rod clad on cast Mg-4.2Zn-1.2Ce-0.7Zr substrate material using Nd:YAG laser. This substrate material is used in aerospace applications. The chosen process parameters were laser power, laser scan speed, and wire feed rate, while the output variable was dilution. The optimum parameter combination for minimum dilution was obtained as laser power of 2250 W, the laser scan speed of 2.5 m/min, and the wire feed rate of 5 m/min. The most significant parameter was found to be the wire feed rate. Dilution was observed to increase with an increase in laser power while decreased with an increase in wire feed rate [37]. The mechanism of formation and solidification of the pool melt while in laser cladding has been reported. The numerical model was built since the experimental determination of the formation mechanism of pool melt could not be achieved. Single-track, multi-track, and multi-layer clad coatings were taken into consideration to determine the convection pattern. The convection occurring inside the clad pool melt was reported to be annular convection. The liquid at the bottom of the melt pool flowed to the melt pool surface, while the liquid on the melt pool surface flowed sideways to the bottom of the melt pool. Hence, the clad coating was formed by the solidification of the liquid metal flowing from the bottom of the pool melt to melt-pool surface [38]. Laser cladding could be employed for the repair of eroded surfaces of materials. One such application was discussed in AISI 420 martensitic stainless steel used in steam turbine blades. During the operating environments, the surface of the steel exposed gets eroded due to impinging water drops. The solution employed here was by laser cladding NiCr and NiCr-TiC powders in order to improve erosion resistance and rigidity of AISI 420 martensitic stainless steel using Nd:YAG laser. Optimization of the process parameters in this process resulted in the combination of laser scan speed of 4 mm/s, laser power of 200 W and powder feed rate of 0.4 g/s. No carbide or any unfavourable phase formation was confirmed in the phase analysis.

The microstructure at the interface corresponded to cellular and planar structures, a columnar dendritic structure at the centre of the clad layer and a coaxial dendritic structure at the top of the clad layer. The mass loss of the substrate with the presence of clad coating was found to decrease three times than that without clad coating, thereby enhancing erosion resistance [39]. The optimal calculations of the powder size were done and found to be in the range of 20 – 120 µm. The lower range was defined by the scattering of the flow of smaller powder particles, while the upper range was defined by the melting incapability of higher mass powder particles. The nozzle with a convergence angle of 64° was determined to be the most productive for which the clad width formed was larger than the diameter of the pool melt by the laser beam. These results could be utilized in the designing of nozzles for cladding powders with any composition [40].

CHARACTERIZATION STUDIES

Microstructural and abrasion wear characteristics of different materials coated with laser cladding have been discussed. Coaxial powder feeding was widely used to deposit tool steel over mild steel using bidirectional scanning pattern, without preheating obtaining crack-free surface and hardness 1000 HV. The portion retaining austenite exhibited better wear resistance when compared to martensite steel coating. Further addition of 20% vanadium carbide greatly enhanced the wear resistance. On the other hand, martensitic tool steel outperformed the austenitic-martensitic tool during single point scratch test. The type of contact between the abrasive and surface explains the difference in wear performance [41]. The coaxial method was also used for cladding in control valve seats using Stellite 6. Hardfacing and plasma transfer methods were compared and found to have better hardness and impact than the PTA technique. Microstructural characterization along the interface of eutectics and carbides revealed microcracks in the PTA-clad specimen, but to a lesser extent for the laser-clad specimen [42]. Fig. (**4a, 4b, 4c** and **4d**) shows the cross-sectional optical micrograph of Stellite 6 cladding [43], Colmonoy 5 cladding [44], Colmonoy 6 cladding [45] and SS420 cladding [46], respectively.

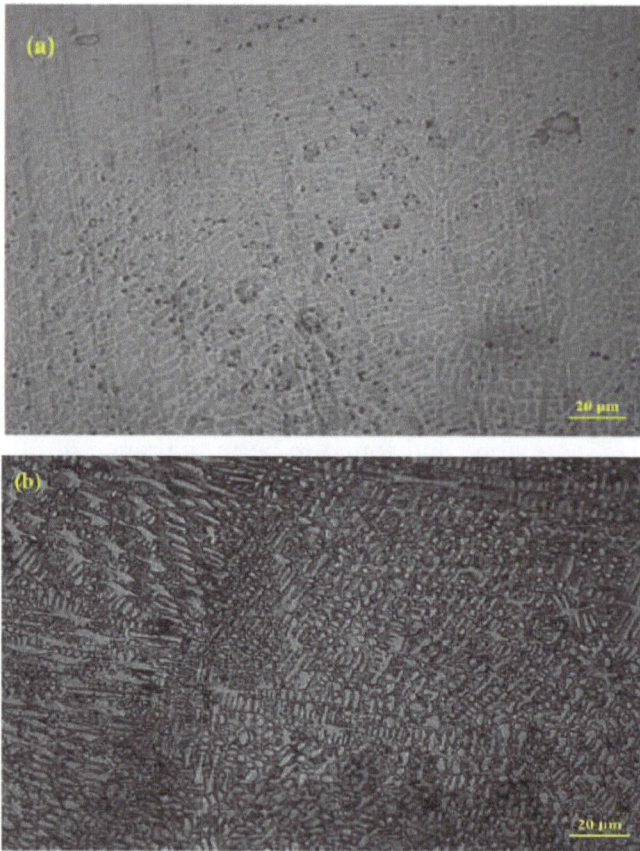

(Fig. 4) contd.....

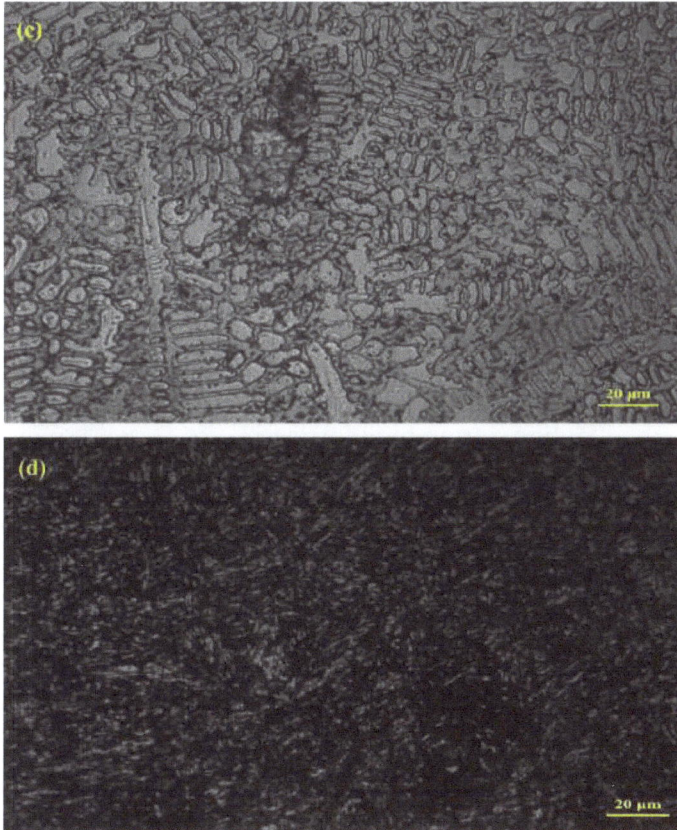

Fig. (4). The cross-sectional optical micrograph of: **(a)** Stellite 6 cladding, **(b)** Colmonoy 5 cladding, **(c)** Colmonoy 6 cladding, **(d)** SS420 cladding.

The laser cladding in different grades of coating exhibited a dendritic structure for all the possible substrates [47, 48]. The laser coating enhanced the hardness and wear behaviour dominated by the combined mechanism of oxidation and abrasion in Stellite 6 layers, while in steel, plastic deformation mechanism also combines with adhesion and abrasion for controlling the wear [49] TiC+δ-(Ti,V)C reinforced Fe based *in situ* multi carbides was coated by laser cladding over AISI 1045 steel substrate. The characteristic feature was examined by SEM, XRD, and dry sliding wear tester. The result revealed the formation of cubic or radial dendrite shaped *in situ* α-Fe Matrix TiC+δ- (Ti, V) C carbides and TiC+δ- (Ti, V) C carbides, and further, the carbide particles were uniformly dispersed in the matrix. δ- (Ti, V) C with dissolved vanadium in TiC structure. A higher microhardness value was obtained by introducing multiple carbides of TiC+δ-(Ti,V)C than TiC [50]. The secondary phase formed in the cermet coating increased the microhardness of Ti-6Al-4V alloy by Ni-ZrB$_2$ coatings. The scanning speed of 1.2 m/min was found to be the most suitable for cermet coating.

Moving from the free surface to substrate, the grain transition changed to fine crystal from mixed cellular grains [51].

A flame resistance coating was prepared using Ti-25V-15Cr-0.2Si to prevent titanium fire. The composition evolution of the transition zone was predicted by the relationship between the dilution ratio and composition. The transformation of α+β bi-phase to a single β phase is visibly characterized using microstructural analysis [52]. Equiaxed-shaped TaC particles were dispersed and cladded with NiCrBSi coating to improve wear resistance. TaC had a better bonding and was interlocked into a matrix rather than being pulled out during the wear process. NiCrBSi combined with Ta experienced abrasion in addition to oxidation. On the other hand, NiCrBSi experienced abrasion with microflaking during low loads and abrasion with oxidation during high loads [53].

The wear resistance of Ti alloy was improved by forming a proper proportion of Ti-Si to improve wear resistance, making the interlayer softer in the substrate and harder in the exterior layer. At 30% Si cladding cracks were developed at the interface and absent in graded coating. Hardness was improved twice and wear resistance four times than that of the substrate [54]. The pulse-based laser cladding of IN718 revealed austenite and interdendritic laves on further refining of microstructure. Nb concentration tends to decrease with an increase in the scanning speed of the laser. The dissolution of Nb with respect to scanning speed increased the microhardness of as-deposited and heat-treated coatings, and further γ''-Ni$_3$Nb precipitation was increased. The wear resistance was improved by heat treatment and scanning speed following the mechanism of fatigue wear associated with abrasive wear [55]. The microstructure of CoNiTi coating on pure Ti observed the formation of BCC dendrites and Ti$_2$Ni interdendritic in the cladding zone. The bonding zone was observed to have fine acicular grains, while the heat-affected zone next to the bonding zone got bulk irregular-shaped grains. Hardness was found to increase 5 times than that of the substrate by solution hardening and second phase hardening of BCC phase and intermetallic components like Ti$_2$Ni and Ti$_2$Co [56]. The use of ultrasonic vibration-assisted laser cladding resulted in a reduction of profile roughness and an increase in coating dilution. Further, it also improved heat and mass transfer resulting in an increased amount of ceramic and homogeneous distribution. This cladding treat increased the wear resistance as well as oxidation resistance in Fe-based composite coating of ceramics [57]. Thick electrical discharge coatings (EDC) deposited wear resistance layer and yielded superior dry sliding wear performance by 4 times against alumina and stellite coating for various loads. The dense region of EDC observed maximum hardness during nanoindentation testing. Microstructural analysis revealed an unusually fine and randomized grain structure compared to the columnar structure of laser coating [58].

In Fe-based coating of Fe-Cr-Mo-Co-C-B-Nb containing Nb powder with 4% mass fraction had good surface properties, corrosion resistance and Glass Forming Ability and with a further increase of Nb mass fraction resulted in the formation of Nb-Mo phase [59]. Non-equiatomic Fe50Mn30Co10Cr10 alloy was prepared, and the effect of boron addition was examined on a 316L substrate. The microstructure was observed to have two crystalline structures: FCC and HCP, with the same chemical composition. HCP structure was obtained because of FCC martensite transformation. The addition of boron led to the formation of the eutectic phase with an interdendritic structure with M2B type Boride. As boride content increased to 5.4%, there was a continuous increase in hardness [60]. The microstructure of dual-layer coating of $CoCrFeMnTi_{0.2}$ high entropy alloy over 15CrMn steel composed of equiaxed crystals of FCC solid solution in the first phase and lamellar eutectic structure surrounded by martensite of the second layer.

This was mainly due to the dilution effect of twin layer deposit and quenching effect. The microhardness of the coating was 3.5 times higher than that of CoCrFeMnNi high entropy alloy and better wear resistance was also reported following abrasive and slight oxidation wear mechanism. However, this coating failed to improve the corrosion resistance compared to CoCrFeMnNi HEA [61]. The damage tolerance ability of aircraft skin was simulated by a double ellipsoid and SCTMA model and stress integrity and fatigue life were calculated by M integral method and Paris equation. The fatigue life was increased by 3 times than that of uncladded aircraft material. Better crack resistance was obtained at a 0-degree cladding angle. The fatigue life was found to improve due to residual compressive stress developed by higher laser power with lower scanning velocity [62]. In $AlCoCrFeNiSi_x$ high-entropy alloy coatings, the effect of silicon composition exhibited body-centered cubic (BCC) structure. Further nanoparticles of AlNi were distributed over the grain, increasing the dislocation density and microhardness with an increase in Si content [8]. Lased cladding based on the paste bound method was used for hard facing alloy in the exact positioning of the laser beam. The technique was applied to Co–Cr–W–B–Si and Co–Ni–Cr–W–B–Si and found to have better wear resistance than plasma spray hard-facing [7]. In cladding of $Al_xCrFeCoNiCu$, as x increased, the corrosion current density first decreased then further increased with percentage addition of x. Still, it could meet the resistivity offered by substrate where $Al_{0.8}CrFe$-CoNiCu experienced better corrosion property. The structure was also found to change from FCC1 to FCC1 + BCC1 and then BCC1 + BCC2 + FCC2 phases with Al addition [63].

High entropy alloy formed by a mixture of metallic and non-metallic compounds led to the development of new material resulting in good refractoriness, corrosion, resistance to irradiation, and wear resistance. Though many cladding techniques like magnetron sputtering, plasma transfer arc, laser-assisted cladding, and electrochemical cladding laser cladding were found to be more beneficial with narrow HAZ, high melting solidification rate, optimum dilution, better metallurgical bonding, and accurate control [64], laser cladding assisted with auxiliary field and electromagnetic ultrasonic compound resulted in better hardness and lower wear volume. This was due to the enhanced convection in the molten pool, inhibiting element segregation and breaking the dendrites to form fine microstructure [65]. In laser cladding of TiN coating in steel 1045, excellent fusion bonding eliminating crack and low dilution had been obtained under the specific energy of range 3.5-5 kW/cm^2. From the bottom of the molten pool, γ nickel growth was found with planar morphology, fine eutectic of γ nickel and (Fe, Cr)$_{23}$C$_6$ in the interdendritic regions [66]. Stellite 6 coating deposition was done on SS316 by varying heat input. The deposition pattern with both presence and absence of buffer layer were obtained crack-free. Hardness increased with the addition of a buffer layer and exhibited a low coefficient of friction as a result of finer grain size and higher microhardness [22]. The ZrB$_2$ coating of Ni-based composite experienced higher microhardness and wear resistance; the presence of the ZrB$_2$ phase further enhanced the hardness of NiCrBSi coating. The ceramic nature of ZrB$_2$ alleviated the adhesive wear, hence increasing the tribological property [67].

DEFECTS AND CONTROL MEASURES

Laser cladding is a surface modification and property improvement process; various defects arise with respect to coating material, medium, thickness, temperature, *etc*. Various papers were reviewed, and their defects and control measures were briefed. Surface alloys with wide-ranging favourable properties can be produced by this laser surface alloying. This method added small quantities of alloying elements to a molten metal pool obtained by localized melting of the surface by the action of the laser, thereby altering the surfaces of inexpensive substrates. Observation of some defects, such as cracks, porosity, surface roughness, compositional inhomogeneity, and excessive dilution was recorded during laser alloying (pulsed Nd:YAG laser) with molybdenum and Ni-Mo on stainless steel and with chromium on mild steel. The techniques used to control such defects have also been outlined [68]. The solidification cracking during laser cladding of Inconel 718 powder on A-286 iron-based hot crack susceptible superalloy was tested. The important processing parameters and geometrical attributes were correlated to identify the required conditions to obtain clads free of cracks. The correlations showed that the foremost macrostructural feature to avoid solidification cracking is the dilution ratio [69, 70].

High Tensile Steel, such as AISI-4340, undergoes extensive wear at the contact areas during operation, resulting in loss of performance and leading to replacement. Laser cladding can be used for the repair of such components, reducing the cost of replacement and being beneficial in terms of environmental concerns. AerMet-100 alloy powder was identified for the repair of AISI4340 steel by laser cladding method and was also observed to generate defect-free clads, which had good compatibility with the substrate. Post-heat treatment of the laser clads at 470°C for 1 hr aided in obtaining homogenous microstructure at HAZ. The microstructure of the as-clad layer revealed lath martensite with retained austenite, while that of the post-heat-treated clad layer revealed slightly reduced retained austenite. An increase in clad hardness was observed after post-heat treatment due to the formation of fine carbides [71]. The phenomenon of dislocation played a vital role in determining the material property of AlCoCrFeNiSix high-entropy alloy coatings. The effect of silicon composition exhibited body-center cubic (BCC) structure. Further nanoparticles of AlNi were distributed over the grain, increasing the dislocation density and microhardness with an increase in Si content [8]. The repair of 340 low-pressure blades in-stream turbine was carried out using laser cladding process. On comparing with other processes, such as open arc hard facing, manual metal or plasma transferred arc welding, cladding with erosion-resistant materials using gas tungsten laser cladding provided better erosion resistance and increased the life cycle [72]. Residual stress generation was controlled by laser cladding techniques in Fe-M--Si-Cr-Ni alloy by stress-induced solid phase transformation. The analysis proved that the generated stress during laser cladding induces austenite to transform into martensite, thereby obtaining coating with low residual stress. This work paved the way for the simulation of stress-strain changes caused by stress-induced solid phase transformation and a new finding in terms of elimination of residual stresses during laser processing [73].

The microstructure of clad $Zr_{65}Al_{7.5}Ni_{10}Cu_{17.5}$ amorphous alloy comprised of white Zr rich particles, greyish Mg-rich eutectics, and grey Mg-rich equiaxed dendrites, with a characteristic sharp interface. A good bonding was confirmed since no micro-voids or pores were identified in the interface. Hunt's CET model and KGT model were chosen to analyse the growth type of the Mg-rich phase in the bonding area, which indicated that the analysis outcome exhibited a good agreement with the experimental results [11]. Laser cladding in the repair of IN738 gas turbine blades was reviewed. The heat treatment procedures associated before and after blade repair were discussed. Laser clad with IN625 powder can be used for repairing IN738 blades in low-stress areas and does not require preheating. Preheating to high temperatures became essential to avoid cracking during laser cladding with high strength material, IN738 powder. Blades repaired with IN738 additive powder were found to be stronger but much more difficult to process than those with IN625 powder [74]. Optimization of process parameters during repair was performed to establish a relationship between laser cladding process parameters (laser power, scan speed, and powder feed rate) and AISI D2 tool steel metallurgical transformations. H13 tool steel powder was deposited on some steel substrates with varying initial metallurgical features (annealed or tempered) by a coaxial laser. Carbide dilution and non-equilibrium phase formation occurred, deteriorating the mechanical properties of the clad layer. The presence of retained austenite at the substrate-coating interface was directly correlated to the cooling rate and resulted in hardness reduction, hence it must be avoided. Tempered substrates involved higher heat accumulation and laser absorption, which led to a bigger HAZ than the annealed substrates [75].

MAPPING FUTURE RESEARCH DIRECTIONS USING CO-OCCURRENCE NETWORK ANALYSIS

To determine the most-studied and future research scopes in the three major casting defects, namely gas and micro-porosity, inclusions, and hot tears, a keyword co-occurrence network (KCON) is constructed following Rajagopal *et al.*'s approach (2017) [76]. In each domain, based on the above-reviewed research papers, a KCON is constructed where each node in the network represents the author-defined keyword, and a link/edge is established between two nodes if a research paper carries both the keywords. The more the number of research papers carrying the two keywords, the more will be the edge weightage between the said keywords. The author-defined keywords are extracted using Bib Excel software, and using the extracted data, the KCON is visualized in Gephi software. Fig. (**5** to **7**) show the KCON of microstructure/mechanical/tribological characterization, defects and their possible remedies, optimization models, and process parameters, respectively, in the Laser Cladding and node sizes are proportional to their Eigenvector scores [76], which determines how central (important) a node is in a network. The nodes concentrated towards the center of the network represent the most-studied areas, and nodes dispersed away from the center denote the least researched area. This distribution of nodes in KCON is achieved by employing the Force Layout algorithm in Gephi.

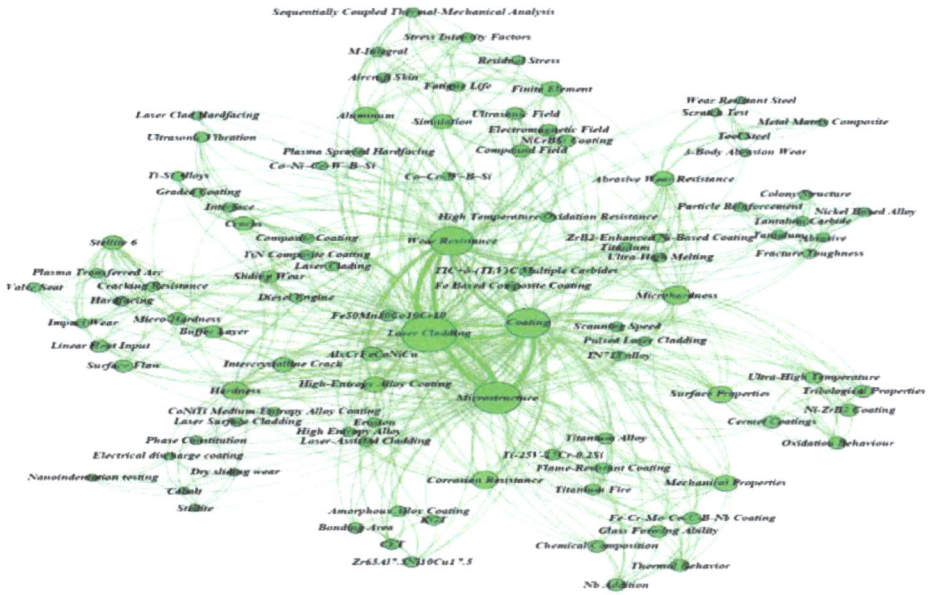

Fig. (5). KCON of microstructure/mechanical/tribological characterization area in laser cladding domain.

From Fig. **(5)**, it can be inferred that among the material characterization studies, micro-hardness, wear resistance, corrosion resistance, and oxidation behavior are predominantly analyzed. The future research directions based on the keywords shifted away from the center are summarized as follows:

- Flame resistance, fatigue life, cracking resistance, impact wear, thermal behavior, and fracture toughness are the least studied mechanical properties.
- CET, KGT, electromagnetic/ultrasonic field testing, 3-body abrasion wear testing, sequentially coupled thermo-mechanical analysis and nano-indentation testing need attention.
- Electrical discharge coupled laser cladding, amorphous alloy coating, cermet coating, hard facing, plasma transferred arc laser cladding, and ultrasonic vibration coupled laser cladding are the emerging variants of laser cladding that need attention.

Similarly, from Fig. **(6)**, most studied research streams in the defects and remedies domain are found as hot crack susceptibility, micro-voids, and compositional inhomogeneities. The future research directions in this domain, as inferred from Fig. **(6)** includes the following:

Fig. (6). KCON of defects and their possible remedies in laser cladding domain.

- Heat accumulation, laser adsorption, excessive dilution, carbide dilution, and non-equilibrium phase formation are the main causes for the defects in the laser cladding, which need to be addressed.
- Refurbishment, post-clad heat treatment, *in-situ* repairs, weld repairs, precipitation strengthening, and solid-solution strengthening are the proposed remedies for laser cladding defects in the literature.
- Defects, such as coating defects, resurface cracks, solidification cracking, and inconsistent bonding-area, are least studied and need attention.

Fig. (7) shows that under the optimization domain, response surface methodology, simulation, and Taguchi's design of experiments are predominantly studied areas. Further, the future research directions in these domains include

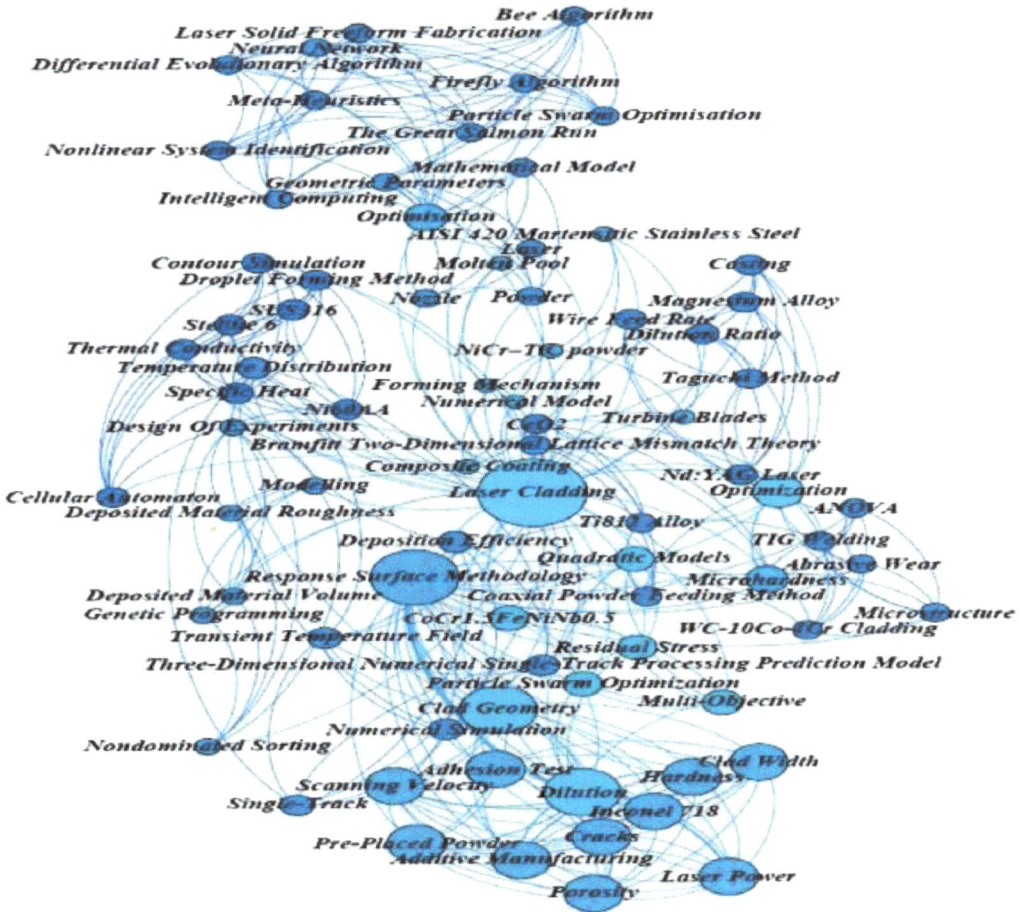

Fig. (7). KCON of optimization techniques in laser cladding domain.

- Studies on the optimization of laser cladding process parameters using artificial intelligence-based optimization techniques, such as neural networks, non-dominating sorting genetic algorithm, particle swarm optimization, cellular automaton, fire-fly, and bee algorithms, are limited and need attention.
- Laser cladding geometric dimensions, droplet forming mechanism, free-form fabrication, and temperature distribution are the least analyzed objective measures in the laser cladding optimization domain.

Fig. (8) shows that under the laser cladding process parameter domain, laser power, scanning velocity, powder feeding rate, and laser focus height are predominantly used input parameters. Further, the future research directions in these domains include:

Fig. (8). KCON of process parameters studied in laser cladding domain.

- Studies on process parameters, such as laser beam profile, traversing speed, melt pool depth, and beam interaction time, are limited and need attention.
- Laser cladding treatment of an inclined substrate, NiCrBSi alloy, and Cr_2Ti are least studied.

CONCLUDING REMARKS

A critical review on laser cladding of different materials is carried out in this chapter. Many special features in laser cladding of different coating onto steel, titanium, and ceramic substrates, showing the applicability in a wide range of materials, are compared. Comparatively, less work was reported for pure aluminium and Al-alloys. Works on thermal modelling of the process were carried out to understand the temperature distribution during the process and to correlate the properties of the clad layers. High flexibility and small area cladding are highly possible in laser cladding. The laser radiation gives better strands like low distortion, phase transformation, minimal diffusion, and low porosity. Optimization of the process parameters has also been investigated to obtain defect-free clads. The major applications of laser cladding are in aircraft engines and automotive industries to improve erosion, wear, oxidation, and corrosion resistance. This gives a broad view of laser cladding.

CONSENT FOR PUBLICATION

Not applicable.

CONFLICT OF INTEREST

The authors declare no conflict of interest.

ACKNOWLEDGEMENTS

None declared.

REFERENCES

[1]   Birger EM, Moskvitin GV, Polyakov AN, Arkhipov VE. Industrial laser cladding: current state and future. Welding International. 2011 Mar 1;25(03):234-43Waynee Penn (2008) Laser Cladding Basics Weld J 47-9.

[2]   Siddiqui AA, Dubey AK. Recent trends in laser cladding and surface alloying. Opt Laser Technol 2021; 134: 106619.
[http://dx.doi.org/10.1016/j.optlastec.2020.106619]

[3]   Liu YH, Li J, Xuan FZ. Fabrication of TiC reinforced Ni based coating by laser cladding. Surf Eng 2012; 28(8): 560-3.
[http://dx.doi.org/10.1179/1743294412Y.0000000026]

[4]   Fu Y, Guo N, Cheng Q, Zhang D, Feng J. *in-situ* formation of laser-cladded layer on Ti-6Al-4 V titanium alloy in underwater environment. Opt Lasers Eng 2020; 131: 106104.
[http://dx.doi.org/10.1016/j.optlaseng.2020.106104]

[5]   Quazi MM, Fazal MA, Haseeb AS, Yusof F, Masjuki HH, Arslan A. A review to the laser cladding of self-lubricating composite coatings. Lasers in Manufacturing and Materials Processing 2016; 3(2): 67-99.
[http://dx.doi.org/10.1007/s40516-016-0025-8]

[6]   Lugscheider E, Bolender H, Krappitz H. Laser cladding of paste bound hardfacing alloys. Surf Eng 1991; 7(4): 341-4.
[http://dx.doi.org/10.1179/sur.1991.7.4.341]

[7]   Liu H, Zhang T, Sun S, Zhang G, Tian X, Chen P. Microstructure and dislocation density of AlCoCrFeNiSix high entropy alloy coatings by laser cladding. Mater Lett 2021; 283: 128746.
[http://dx.doi.org/10.1016/j.matlet.2020.128746]

[8]   Wang AH, Xie CS, Nie JH. Microstructural characteristics of iron based alloy laser clad on Al–Si alloy. Mater Sci Technol 1999; 15(8): 957-64.
[http://dx.doi.org/10.1179/026708399101506643]

[9]   Song B, Yu T, Jiang X, Xi W, Lin X. Development mechanism and solidification morphology of molten pool generated by laser cladding. Int J Therm Sci 2021; 159: 106579.
[http://dx.doi.org/10.1016/j.ijthermalsci.2020.106579]

[10] Su Y, Yue T. Microstructures of the bonding area in laser cladded Zr-based amorphous alloy coating on magnesium. Mater Today Commun 2020; 25: 101715.
[http://dx.doi.org/10.1016/j.mtcomm.2020.101715]

[11] Kripalani K, Jain P. Comprehensive study of laser cladding by nitinol wire. Mater Today Proc 2020.

[12] Kaiming W, Yulong L, Hanguang F, Yongping L, Zhenqing S, Pengfei M. A study of laser cladding NiCrBSi/Mo composite coatings. Surf Eng 2018; 34(4): 267-75.
[http://dx.doi.org/10.1080/02670844.2016.1259096]

[13] Qi K, Yang Y, Hu G, Lu X, Li J. Thermal expansion control of composite coatings on 42CrMo by laser cladding. Surf Coat Tech 2020; 397: 125983.
[http://dx.doi.org/10.1016/j.surfcoat.2020.125983]

[14] Liu J, Yu H, Chen C, Weng F, Dai J. Research and development status of laser cladding on magnesium alloys: A review. Opt Lasers Eng 2017; 93: 195-210.
[http://dx.doi.org/10.1016/j.optlaseng.2017.02.007]

[15] Vandehaar E, Malian PA, Baldwin M. Laser cladding of thermal barrier coatings. Surf Eng 1988; 4(2): 159-72.
[http://dx.doi.org/10.1179/sur.1988.4.2.159]

[16] Jeyaprakash N, Yang C-H, Sivasankaran S. Laser cladding process of Cobalt and Nickel based hard-micron-layers on 316L-stainless-steel-substrate. Mater Manuf Process 2020; 35(2): 142-51.
[http://dx.doi.org/10.1080/10426914.2019.1692354]

[17] Shu D, Li Z, Yao C, Li D, Dai Z. *In situ* synthesised WC reinforced nickel coating by laser cladding. Surf Eng 2018; 34(4): 276-82.
[http://dx.doi.org/10.1080/02670844.2017.1320057]

[18] Wargulski DR, Nowak T, Thiele M, *et al.* Quality management of laser cladding processes for additive manufacturing by new methods of visualisation and evaluation of thermographic data. Quant Infrared Thermogr J 2020; 17(1): 1-2.
[http://dx.doi.org/10.1080/17686733.2019.1592392]

[19] Cottam R, Brandt M. Laser Cladding of Ti-6Al-4 V Powder on Ti-6Al-4 V Substrate: Effect of Laser Cladding Parameters on Microstructure. Phys Procedia 2011; 12: 323-9.
[http://dx.doi.org/10.1016/j.phpro.2011.03.041]

[20] Wang Q, Shi J, Zhang L, Tsutsumi S, Feng J, Ma N. Impacts of laser cladding residual stress and material properties of functionally graded layers on titanium alloy sheet. Addit Manuf 2020; 35: 101303.
[http://dx.doi.org/10.1016/j.addma.2020.101303]

[21] Thawari N, Gullipalli C, Katiyar JK, Gupta TV. Influence of buffer layer on surface and tribomechanical properties of laser cladded Stellite 6. Mater Sci Eng B 2021; 263: 114799.
[http://dx.doi.org/10.1016/j.mseb.2020.114799]

[22] Weng F, Yu H, Chen C, *et al.* Effect of process parameters on the microstructure evolution and wear property of the laser cladding coatings on Ti-6Al-4V alloy. J Alloys Compd 2017; 692: 989-96.
[http://dx.doi.org/10.1016/j.jallcom.2016.09.071]

[23] Jiang GY, Liu YP, Xie JL, Wang WC. Mechanical and corrosion resistance of laser cladding Ni-based alloy of steel plate under variable defocusing. Optik (Stuttg) 2020; 224: 165464.
[http://dx.doi.org/10.1016/j.ijleo.2020.165464]

[24] Li M, Zhang Q, Han B, Song L, Li J, Zhang S. Effects of ultrasonic impact treatment on structures and properties of laser cladding Al0. 5CoCrFeMnNi high entropy alloy coatings. Mater Chem Phys 2021; 258: 123850.
[http://dx.doi.org/10.1016/j.matchemphys.2020.123850]

[25] Xu PY, Liu YC, Yi P, Fan CF, Li CK. Effect of ultrarapid cooling on microstructure of laser cladding IN718 coating. Surf Eng 2013; 29: 414-8.
[http://dx.doi.org/10.1179/1743294413Y.0000000142]

[26] Kathuria YP. Role of beam interaction time in laser cladding process. Mater Sci Technol 2001; 17(11): 1451-4.
[http://dx.doi.org/10.1179/026708301101509458]

[27] Zhao Y, Guan C, Chen L, Sun J, Yu T. Effect of process parameters on the cladding track geometry fabricated by laser cladding. Optik (Stuttg) 2020; 223: 165447.
[http://dx.doi.org/10.1016/j.ijleo.2020.165447]

[28] Nair A, Ramji V, Raj RD, Veeramani R. Laser cladding of Stellite 6 on EN8 steel–A fuzzy modelling approach. Mater Today Proc 2020.

[29]   Tanigawa D, Abe N, Tsukamoto M, *et al.* Effect of laser path overlap on surface roughness and hardness of layer in laser cladding. Sci Technol Weld Join 2015; 20(7): 601-6.
[http://dx.doi.org/10.1179/1362171815Y.0000000044]

[30]   Kai-ming W, Han-guang F, Yu-long L, Yong-ping L, Shi-zhong W, Zhen-qing S. Effect of power on microstructure and properties of laser cladding NiCrBSi composite coating. Transactions of the IMF. 328-6.
[http://dx.doi.org/10.1080/00202967.2017.1355640]

[31]   Han T, Xiao M, Zhang Y, Shen Y. Laser cladding Ni-Ti-Cr alloy coatings with different process parameters. Mater Manuf Process 2019; 34(15): 1710-8.
[http://dx.doi.org/10.1080/10426914.2019.1686521]

[32]   Javid Y. Multi-response optimization in laser cladding process of WC powder on Inconel 718. CIRO J Manuf Sci Technol 2020; 31: 406-17.
[http://dx.doi.org/10.1016/j.cirpj.2020.07.003]

[33]   Lotfy K, Gabr ME. Response of a semiconducting infinite medium under two temperature theory with photothermal excitation due to laser pulses. Opt Laser Technol 2017; 97: 198-208.
[http://dx.doi.org/10.1016/j.optlastec.2017.06.021]

[34]   Ma M, Xiong W, Lian Y, Han D, Zhao C, Zhang J. Modeling and optimization for laser cladding *via* multi-objective quantum-behaved particle swarm optimization algorithm. Surf Coat Tech 2020; 381: 125129.
[http://dx.doi.org/10.1016/j.surfcoat.2019.125129]

[35]   Cao X, Xiao M, Jahazi M, Fournier J, Alain M. Optimization of processing parameters during laser cladding of ZE41A-T5 magnesium alloy castings using Taguchi method. Mater Manuf Process 2008; 23(4): 413-8.
[http://dx.doi.org/10.1080/10426910801940391]

[36]   Song B, Yu T, Jiang X, Xi W. Numerical model of transient convection pattern and forming mechanism of molten pool in laser cladding. Numer Heat Transf A 2019; 75(12): 855-73.
[http://dx.doi.org/10.1080/10407782.2019.1608777]

[37]   Saeedi R, Razavi RS, Bakhshi SR, Erfanmanesh M, Bani AA. Optimization and characterization of laser cladding of NiCr and NiCr–TiC composite coatings on AISI 420 stainless steel. Ceram Int 2021; 47(3): 4097-110.
[http://dx.doi.org/10.1016/j.ceramint.2020.09.284]

[38]   Grigoryants AG, Tretyakov RS, Shiganov IN, Stavertiy AY. Optimization of the shape of nozzles for coaxial laser cladding. Welding International 2015; 29(8): 639-42.
[http://dx.doi.org/10.1080/01431161.2014.967043]

[39]   Tuominen J, Näkki J, Pajukoski H, Hyvärinen L, Vuoristo P. Microstructural and abrasion wear characteristics of laser-clad tool steel coatings. Surf Eng 2016; 32(12): 923-33.
[http://dx.doi.org/10.1080/02670844.2016.1180496]

[40]   Chang SS, Wu HC, Chen C. Impact wear resistance of stellite 6 hardfaced valve seats with laser cladding. Mater Manuf Process 2008; 23(7): 708-13.
[http://dx.doi.org/10.1080/10426910802317102]

[41]   Jeyaprakash N, Yang CH, Tseng SP. Wear Tribo-performances of laser cladding Colmonoy-6 and Stellite-6 Micron layers on stainless steel 304 using Yb: YAG disk laser. Met Mater Int 2019; 13: 1-4.

[42]   Jeyaprakash N, Yang CH, Ramkumar KR, Sui GZ. Comparison of microstructure, mechanical and wear behaviour of laser cladded stainless steel 410 substrate using stainless steel 420 and Colmonoy 5 particles. J Iron Steel Res Int 2020; 27(12): 1446-55.
[http://dx.doi.org/10.1007/s42243-020-00447-4]

[43]   Jeyaprakash N, Yang CH, Ramkumar KR. Microstructure and wear resistance of laser cladded Inconel 625 and Colmonoy 6 depositions on Inconel 625 substrate. Appl Phys, A Mater Sci Process 2020; 126:

1-1.
[http://dx.doi.org/10.1007/s00339-020-03637-9]

[44]   Jeyaprakash N, Yang CH. Microstructure and Wear Behaviour of SS420 Micron Layers on Ti–6Al–4V Substrate Using Laser Cladding Process. Trans Indian Inst Met 2020; 12: 1-7.
[http://dx.doi.org/10.1007/s12666-020-01927-7]

[45]   Jeyaprakash N, Yang CH. Comparative study of NiCrFeMoNb/FeCrMoVC laser cladding process on nickel-based superalloy. Mater Manuf Process 2020; 35(12): 1383-91.
[http://dx.doi.org/10.1080/10426914.2020.1779933]

[46]   Jeyaprakash N, Yang CH, Tseng SP. Characterization and tribological evaluation of NiCrMoNb and NiCrBSiC laser cladding on near-α titanium alloy. Int J Adv Manuf Technol 2020; 106(5): 2347-61.
[http://dx.doi.org/10.1007/s00170-019-04755-2]

[47]   Jeyaprakash N, Yang C-H, Sivasankaran S. Formation of FeCrMoVC Layers on AA6061 by Laser Cladding Process: Microstructure and Wear Characteristics. Trans Indian Inst Met 2020; 73(6): 1611-7.
[http://dx.doi.org/10.1007/s12666-020-01942-8]

[48]   Wang XH, Zhang M, Du BS, Qu SY, Zou ZD. Microstructure and wear properties of *in situ* multiple carbides reinforced Fe based surface composite coating produced by laser cladding. Mater Sci Technol 2010; 26(8): 935-9.
[http://dx.doi.org/10.1179/026708309X12459430509094]

[49]   Farotade GA, Adesina OS. foluso Ogunbiyi O, tobi Adesina O, Kazeem RA, Adeniran AA. Microstructural characterization and surface properties of laser clad Ni-ZrB2 coatings on Ti-6Al-4V alloy. Mater Today Proc 2020.

[50]   Zhang F, Qiu Y, Hu T, Clare AT, Li Y, Zhang LC. Microstructures and mechanical behavior of beta-type Ti-25V-15Cr-0.2 Si titanium alloy coating by laser cladding. Mater Sci Eng A 2020; 796: 140063.
[http://dx.doi.org/10.1016/j.msea.2020.140063]

[51]   Yu T, Deng QL, Zheng JF, Dong G, Yang JG. Microstructure and wear behaviour of laser clad NiCrBSi+ Ta composite coating. Surf Eng 2012; 28(5): 357-63.
[http://dx.doi.org/10.1179/1743294411Y.0000000094]

[52]   Yue TM, Zhang H. Microstructure and wear resistance of compositionally graded Ti–Si laser-clad coatings. Materials Research Innovations 2015; 19((sup9)): S9-9.
[http://dx.doi.org/10.1179/1432891715Z.0000000001905]

[53]   Zhang Y, Yang L, Dai J, *et al.* Microstructure and mechanical properties of pulsed laser cladded IN718 alloy coating. Surf Eng 2018; 34(4): 259-66.
[http://dx.doi.org/10.1080/02670844.2016.1200847]

[54]   Xiang K, Chai L, Wang Y, *et al.* Microstructural characteristics and hardness of CoNiTi medium-entropy alloy coating on pure Ti substrate prepared by pulsed laser cladding. J Alloys Compd 2020; 849: 156704.
[http://dx.doi.org/10.1016/j.jallcom.2020.156704]

[55]   Zhang M, Zhao GL, Wang XH, Liu SS, Ying WL. Microstructure evolution and properties of *in-situ* ceramic particles reinforced Fe-based composite coating produced by ultrasonic vibration assisted laser cladding processing. Surf Coat Tech 2020; 403: 126445.
[http://dx.doi.org/10.1016/j.surfcoat.2020.126445]

[56]   Murray JW, Ahmed N, Yuzawa T, *et al.* Dry-sliding wear and hardness of thick electrical discharge coatings and laser clads. Tribol Int 2020; 150: 106392.
[http://dx.doi.org/10.1016/j.triboint.2020.106392]

[57]   Wang HZ, Cheng YH, Yang JY, Wang QQ. Microstructure and properties of laser clad Fe-based amorphous alloy coatings containing Nb powder. J Non-Cryst Solids 2020; 550: 120351.

[http://dx.doi.org/10.1016/j.jnoncrysol.2020.120351]

[58]   Aguilar-Hurtado JY, Vargas-Uscategui A, Paredes-Gil K, Palma-Hillerns R, Tobar MJ, Amado JM. Boron addition in a non-equiatomic Fe50Mn30Co10Cr10 alloy manufactured by laser cladding: Microstructure and wear abrasive resistance. Appl Surf Sci 2020; 515: 146084.
[http://dx.doi.org/10.1016/j.apsusc.2020.146084]

[59]   Liu H, Li X, Liu J, Gao W, Du X, Hao J. Microstructural evolution and properties of dual-layer CoCrFeMnTi0. 2 high-entropy alloy coating fabricated by laser cladding. Opt Laser Technol 2021; 134: 106646.
[http://dx.doi.org/10.1016/j.optlastec.2020.106646]

[60]   Song M, Wu L, Liu J, Hu Y. Effects of laser cladding on crack resistance improvement for aluminum alloy used in aircraft skin. Opt Laser Technol 2021; 133: 106531.
[http://dx.doi.org/10.1016/j.optlastec.2020.106531]

[61]   Li Y, Shi Y. Microhardness, wear resistance, and corrosion resistance of AlxCrFeCoNiCu high-entropy alloy coatings on aluminum by laser cladding. Opt Laser Technol 2021; 134: 106632.
[http://dx.doi.org/10.1016/j.optlastec.2020.106632]

[62]   Menghani J, Vyas A, Patel P, Natu H, More S. Wear, erosion and corrosion behavior of laser cladded high entropy alloy coatings–A review. Mater Today Proc 2020.
[http://dx.doi.org/10.1016/j.matpr.2020.08.763]

[63]   Hu G, Yang Y, Sun R, Qi K, Lu X, Li J. Microstructure and properties of laser cladding NiCrBSi coating assisted by electromagnetic-ultrasonic compound field. Surf Coat Tech 2020; 404: 126469.
[http://dx.doi.org/10.1016/j.surfcoat.2020.126469]

[64]   Lei TC, Ouyang JH, Pei YT, Zhou Y. Microstructure and sliding wear properties of laser clad TiN reinforced composite coating. Surf Eng 1996; 12(1): 55-60.
[http://dx.doi.org/10.1179/sur.1996.12.1.55]

[65]   Guo C, Zhou J, Zhao J, Chen J. Effect of ZrB2 on the microstructure and wear resistance of Ni-based composite coating produced on pure Ti by laser cladding. Tribol Trans 2010; 54(1): 80-6.
[http://dx.doi.org/10.1080/10402004.2010.519860]

[66]   Goswami GL, Kumar D, Grover AK, Pappachan AL, Totlani MK. Control of defects during laser surface alloying. Surf Eng 1999; 15(1): 65-70.
[http://dx.doi.org/10.1179/026708499322911674]

[67]   Alizadeh-Sh M, Marashi SP, Ranjbarnodeh E, Shoja-Razavi R. Laser cladding of Inconel 718 powder on a non-weldable substrate: Clad bead geometry-solidification cracking relationship. J Manuf Process 2020; 56: 54-62.
[http://dx.doi.org/10.1016/j.jmapro.2020.04.045]

[68]   Alizadeh-Sh M, Marashi SP, Ranjbarnodeh E, Shoja-Razavi R, Oliveira JP. Prediction of solidification cracking by an empirical-statistical analysis for laser cladding of Inconel 718 powder on a non-weldable substrate. Opt Laser Technol 2020; 128: 106244.
[http://dx.doi.org/10.1016/j.optlastec.2020.106244]

[69]   Aditya YN, Srichandra TD, Tak M, Padmanabham G. To study the laser cladding of ultra high strength AerMet-100 alloy powder on AISI-4340 steel for repair and refurbishment. Mater Today Proc 2020.

[70]   Brandt M, Sun S, Alam N, Bendeich P, Bishop A. Laser cladding repair of turbine blades in power plants: from research to commercialisation. International Heat Treatment and Surface Engineering 2009; 3(3): 105-14.
[http://dx.doi.org/10.1179/174951409X12542264513843]

[71]   Tian J, Xu P, Liu Q. Effects of stress-induced solid phase transformations on residual stress in laser cladding a Fe-Mn-Si-Cr-Ni alloy coating. Mater Des 2020; 193: 108824.
[http://dx.doi.org/10.1016/j.matdes.2020.108824]

[72]   Chen C, Wu HC, Chiang MF. Laser cladding in repair of IN738 turbine blades. International Heat Treatment and Surface Engineering 2008; 2(3-4): 140-6.
[http://dx.doi.org/10.1179/174951508X446484]

[73]   Candel JJ, Amigó V, Ramos JA, Busquets D. Problems in laser repair cladding a surface AISI D2 heat-treated tool steel. Welding International 2013; 27(1): 10-7.
[http://dx.doi.org/10.1080/09507116.2011.592707]

[74]   Rajagopal V, Venkatesan SP, Goh M. Decision-making models for supply chain risk mitigation: A review. Comput Ind Eng 2017; 113: 646-82.
[http://dx.doi.org/10.1016/j.cie.2017.09.043]

CHAPTER 9

Strengthening of Tribological Properties by Laser Texturing: Impact of Process Variables and Applications – Review and Future Trends

Varthini Rajagopal[1], **R. Dinesh Kumar**[2,*] and **K. Ganesa Balamurugan**[3]

[1] Department of Mechanical Engineering, Government College of Engineering Srirangam, Tiruchirappalli- 620012, Tamil Nadu, India

[2] Department of Mechanical Engineering, Indian Institute of Technology, Guwahati- 781039, Assam, India

[3] Department of Mechanical Engineering, IFET College of Engineering- Villupuram- 605108, Tamil Nadu, India

Abstract: Recent research has focused on enhancing the tribological and wettability characteristics of materials through surface texturing, which involves the formation of specific patterns on the material surface through abrasive blasting, reactive-ion etching, lithography, and mechanical machining. One such competitive technique is laser texturing, which uses high-energy pulses to remove and etch material by direct absorption of laser energy followed by rapid melting and vaporizing. This chapter discusses the influences of laser texturing process parameters, such as laser wavelength, laser focusing technique, laser energy, pulse repetition rate, pulse duration, focal distance, direct/in-direct processing, dimple density, and geometry of texturing on the tribological properties. Next, the limitations in laser texturing and the post-processing techniques to rectify those challenges are reviewed. Further, the recent advances of laser texturing in bio-medical applications for prolonging the service life of bio-implants are detailed.

Keywords: Automobile, Bio-implants, Coating adhesion strength, Contact angle measurement, Corrosion, 2D and 3D topography, Friction, Hydrophilicity, Hydrophobicity, Laser fluence, Laser manufacturing, Laser scan speed, Laser surface modifications, Laser surface texturing, Laser texturing, LST modelling, Mechanical anchoring, Micro and nano texturing, Number of shots, Post-Processing, Surface modification of materials, Tribology, Wear.

---

* **Address correspondence to R. Dinesh Kumar:** Department of Mechanical Engineering, Indian Institute of Technology- Guwahati, Assam, India; Tel: +91-9865138376; Email: dineshrd453@gmail.com

Jeyaprakash Natarajan and Che-Hua Yang (Eds.)

INTRODUCTION

Wear and tear result due to the friction or the interaction of the moving objects. Friction and wear are two major concerns in various engineering components that decide the product's life cycle. To obtain maximum energy efficiency and increase the durability of the component, it is always required to have curb friction and wear between the mating surface [1]. The very first approach to surface texturing was successfully tested by developing a line interface for a piston ring of a diesel engine in 1940 [2]. The laser surface texturing (LST) and shock peening are shown in Figs. (1) and (2).

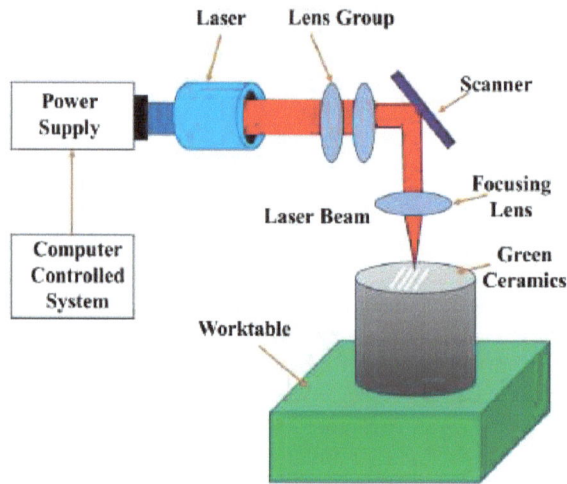

Fig. (1). Schematic representation of laser surface texturing.

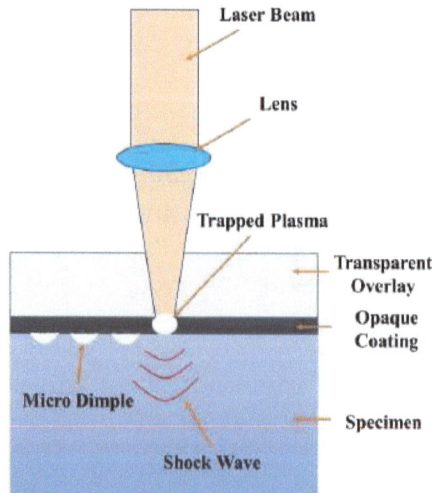

Fig. (2). Schematic representation of laser peening.

Later in 1960, a theory was proposed on micro lubrication by introducing protrusion to develop tribological performance [3]. A new trend of introducing micro dimples was found to result in reduced friction and wear resistance. The major reason for this is found to be dimples with a closed shape that increase the hydrodynamic pressure. The surface integrity of the component also plays a vital role in other characters like crack initiation, corrosion resistance, wettability, *etc*. Enormous efforts have been taken by researchers over the past few centuries in controlling the tribological performance in different materials, whose surfaces mate with each other and slide to operate with varying surface textures [4, 5]. This surface texturing has been identified as the best approach to change the interaction or contact between the surfaces for better friction control as well as lubrication to improve the wear resistance behavior of the engineering materials [6, 7]. To date, there is plenty of research carried out on surface texturing both by numerical and experimental approaches to improve the surface property of the material. The various methods developed for improving surface integrity are micro-milling, electrochemical machining, ion beam etching, hot embossing, mechanical texturing, lithography, and laser surface texturing [8]. The various shapes of texturing are shown in Fig. **(3)**.

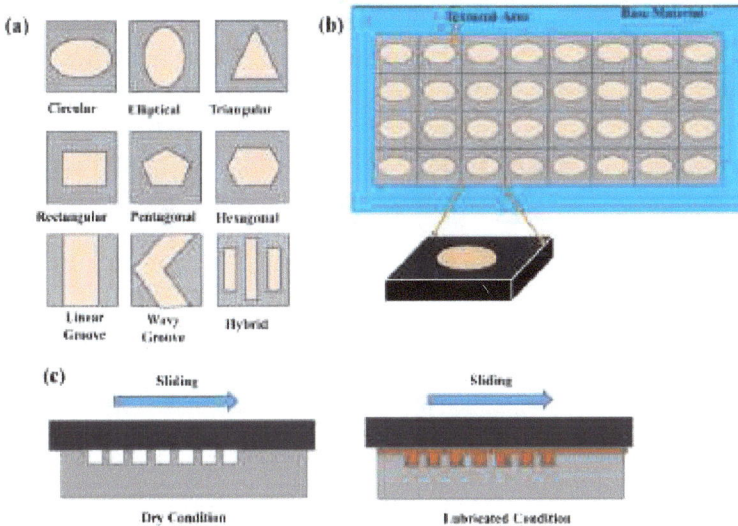

Fig. (3). Schematic illustration of **(a)** different kinds of texture feature shapes, **(b)** texturing parameters during the LST process, and **(c)** the effect of texture design on the contact condition during dry and lubricated sliding.

The vibration mechanism is used for creating texturing in ultrasonic texturing at high frequency, and low magnitude in the slurry environment over the workpiece causes erosion in the material and is found suitable for brittle material. In micro EDM, the erosion or texture formation is through the dielectric fluid interaction between tool and workpiece, and this cannot be used for electrically insulated material. The other texturing methods have got one or other drawbacks with the quality or material possibility, *etc.*, [8]. Even though there are a lot of methods that are used in surface modification, laser surface texturing is found to be advantageous over all other techniques due to its salient features like the ability to control, speed, and higher efficiency compared to other surface texturing techniques.

This is also found to be environmentally friendly and can fabricate surfaces with higher accuracy [9, 10]. Laser texturing has been used in most applications like magnetic storage devices, thrust bearing, mechanical seals, piston rings due to its high tribological properties. Apart from that, it is also being used for various materials like polymers, metals, and ceramics to have better tribological performances [7]. Many researchers, beginning with Etsion and Bonse, have studied the effect of laser texturing.. In recent years, many processes based on laser texturing have been discussed, like laser shock processing and laser interface, introducing alternative methodology and expanding the necessity of LS in tribology [11, 12]. Surface texture is considered to be an important approach to overcome the striction and adhesion in micro-electrical devices [13]. To capture wear debris, micro reservoirs are provided by surface texturing to enhance micro traps or lubricant retention. The texturing dimension optimization is mostly carried out by trial and error method, and many researchers are recently trying to simulate or model the process to optimize the geometry of texturing to achieve better performance [14]. Laser surface texturing is capable of producing micro dimples that serve as micro hydrodynamic bearing with full or mixed lubrication; during starved conditions, it acts like a micro reservoir for lubricant and during dry sliding, it acts like a micro trap for wear debris.

This is mostly used in thrust bearing, mechanical seals, and piston rings [15]. Many experiments are performed using short-pulse laser for microstructure formation, however, one research used a long-pulse laser to create dimples and bump shapes, and found surface tension to be the most important factor for bump shape. Underwater condition is found to be better than air medium [16]. The efficiency of the solar panel is improved by surface texturing using an ArF excimer laser. The surface integrity produced by ArF increases efficiency drastically [17]. The impact of laser shock peening in the surface waviness, machining texture, and direction, surface roughness was analyzed and it was found that this process has more impact on surface waviness and least on surface roughness [11]. A comparative study on piston rings with conventional and laser textured cylinders found a 4% increase in efficiency by reducing fuel consumption without affecting the exhaust level [18]. LST is found to be a better alternative for dental implantology as it takes lesser time for stem cell growth and produces bone tissue along with the implant by bone tissue engineering [19]. This chapter discusses in detail the laser surface texturing and the impact of various process variables, such as friction and wear, friction and lubrication, temperature, a texture design, adhesion and wettability. Apart from that, the application of laser texturing in different fields and a few discussions on modeling are also summarised. The end portion describes the possible future research area to explore in laser surface texturing based on a thorough review of current literature, and previous literature review reveals many research gaps in LST, which needs to be fully explored.

EFFECT OF PROCESS PARAMETERS ON ADHESION

Materials surfaces are coated with special-purpose materials to enhance their surface properties. However, the objective of this surface coating can be achieved with the proper adhesiveness of the coating with the substrate.

LST has been identified as a suitable technique to improve the adhesion of coating with the substrate. He *et al.* [20] have performed the LST on CMG glass to improve the adhesion of the electroless copper-plated layer. A KrF excimer laser was used for the texturing with the process parameters of energy density, shots per area, and pulse repetition. The micro texturing was machined on the surface of the substrate in a 1 mm circular mask and each parameter was tested per grid. The mask overlapping pattern was also considered as a parameter for the investigation.

The formation of the textures depends on the input laser parameters. Increasing the laser process parameter values improves the homogeneity of the machining [21, 22]. The surface topography constitutes the large-scale grid-like structures and small-scale micro-roughness components. The large-scale grid-like structures were influenced by the masking pattern, and the micro-roughness was influenced by the laser parameters. Moreover, the LST has significantly improved the adhesion of the electroless copper layer on the glass substrate. This investigation also suggested that post machining annealing should not be carried out on the glass substrate due to the formation of microcracks.

Challenges in the machining of aluminum and its alloys are the adherence of the machined chips to the cutting tools. This chip adherence to the tool edges might lead to tool breakage. The effect of chip formation due to texture is shown in Fig. (4). Enomoto and Sugihara [23] have overcome this issue by texturing the tool edges with ;aser. A cemented carbide tool with DLC coating was used. A femtosecond titanium-sapphire-based laser system was used for the fabrication of micro/nano textures on the rake faces of the cutting tool with 100-150 nm depth and 700 nm apart. Three types of tools were investigated in the study; (1) DLC coated tools with grooves parallel to the cutting edges, (2) DLC coated tools with grooves orthogonal to cutting edges, and (3) Un-textured DCL tools. The cutting tool performances were evaluated through SEM and EDX. The results revealed that the anti-adhesive property was improved in the micro/nano-textured tools. Especially the tools textured parallel to the cutting edges show an excellent improvement. The texture direction greatly influences the anti-adhesion nature of the chip with the tool face [23, 24]. The orthogonally textured surfaces facilitate the adhesion of chips during their flow on the tool face and break the cutting fluid film. However, in the case of parallel grooved textures, the tool face avoids the chip adhesion due to intermittent contact with it and flush out the chips by by allowing the cutting fluid through the textured grooves. Romoli *et al.* [25] have used LST to enhance the adhesive strength of butt-jointed aluminum alloy components. LST was performed using Nd:YVO4 laser system with the laser power between 12-18 W and the tangential scan speed between 100-1000 mm/s. The textures are groove-shaped with a hatch distance of 0.035-0.200 mm. Finally, the environmental impact of relative humidity and temperature are investigated using contact start-stop testing. The results showed that the temperature is inversely proportional to stiction with respect to increased humidity. This environmental change also affects the lubricant properties [26]. L. Rapoport *et al.* had applied LST to enhance the adhesion of the CdZnSe sub-layer on the ground steel. Over the CdZnSe sub-layer, MoS_2 film was burnished to improve the wear life of the ground steel. The CdZnSe sub-layer provides better adhesion to MoS_2 film, and in turn, the texturing had increased the adhesion between the CdZnSe sub-layer and ground steel. The solid lubrication film filled the microcavities of the texture and resisted the shear forces. R. Kromer *et al.* [27] had investigated the effect of texturing on the coating adhesion strength extensively. Cold spraying and plasma spraying technologies were adopted. For cold spraying, aluminium (7000 series), magnesium alloy (RZ5), and Al-SiC (20 vol%) composites were used as the substrates. Likewise, the coating materials were aluminium (6000), magnesium (RZ5), and Al-SiC composites. A ZrO_2-7Y2O$_3$-1.7HfO$_2$ thermal barrier coating was applied on the AM1 single crystal superalloy substrate by plasma spraying. Textures were created on the respective substrates before the application of coating. The adhesion strength of the coating with the substrate depends on the adhesion area. The adhesion area ratio, R, determines the area covered by the textures, which enhances the adhesion. R. Kromer *et al.* have provided an equation to represent this adhesion area ratio, and this equation is given in eq. (1).

R= Adhesion area/Plane area (1)   **(1)**

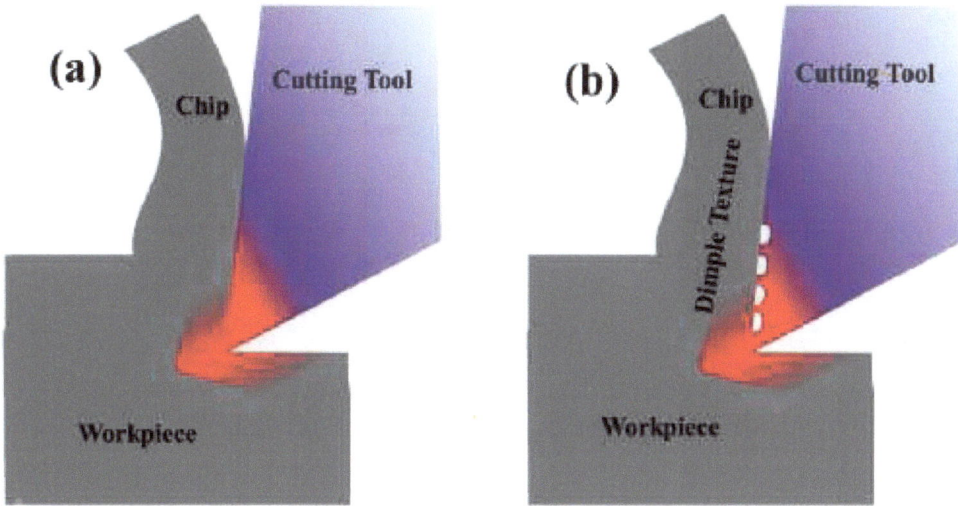

Fig. (4). Effect of chip formation on chip-tool contact: **(a)** without texture **(b)** with dimple texture.

The mechanism of coating adhesion in the textures is discussed in detail in this article. Mechanisms of coating adhesion depend on the coating methods. In cold spraying, both the particles and the substrates experience plastic deformations during the impact. Particle adhesion depends on the concavity of the cavities. LST was concerned with the creation of these concave cavities on the substrates. Plastic deformation was lower and the temperature was higher at the cavities. An increase in temperature promoted metallurgical bonding. In plasma sprayed coating, molten particles were deposited over the surface of the substrate and started spreading. The speed of spreading was restricted by the rapid solidification of the bottom zone of the droplets. When the molted particles were deposited on the textures, the molten particles started filling the cavities of the textures.

The hole filling was improved in the texture cavities wall having a 90° angle at the top surface of the substrate. A filled droplet is solidified inside the cavities of the textures and forms a metallurgical bonding with them. Textures provided mechanical anchoring, which increases the failure energies. Fig. (**5**) shows the schematic representation of this mechanism. Thus, the textures significantly improve the adhesion strength between the coating and the substrate. Therefore, during the relative motion between any counter surfaces, the textured surfaces mechanically anchored the coating materials and only the top portion of them would be fractured instead of completely peeling off of the coating from the substrate. Fig. (**6**) schematically explains this concept. This gives the detail about how texturing helps in metal transfer for better efficiency. The optimal process parameter for promoting the adhesive behavior of the material is listed in Table **1**.

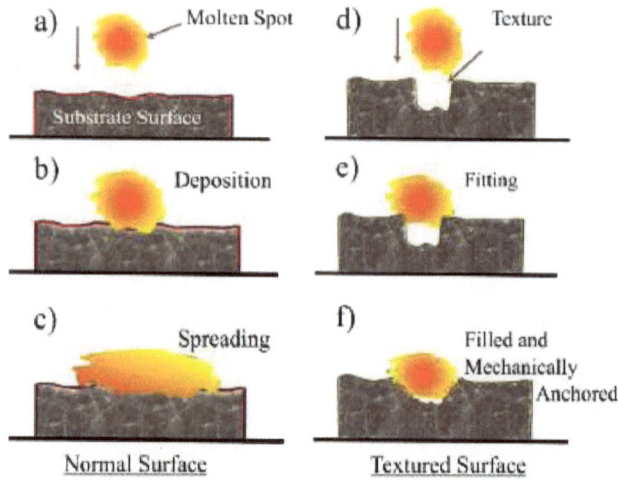

Fig. (5). Schematic representation of droplet spreading mechanism on un-textured and textured surfaces.

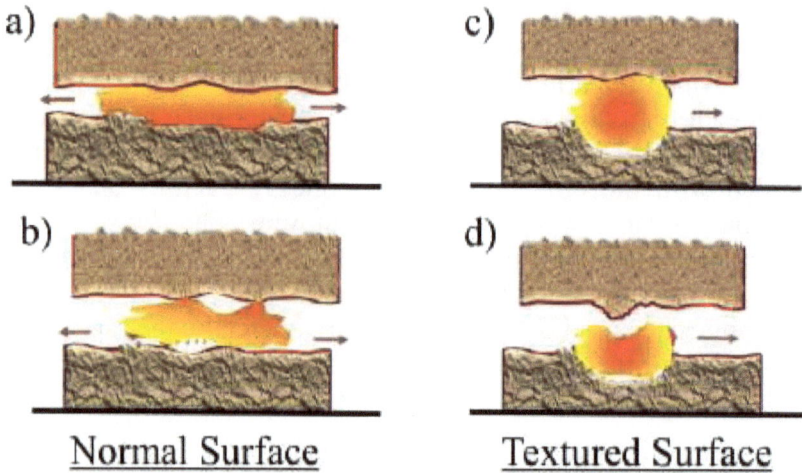

Fig. (6). Coating failure mechanism in normal and textured surfaces; a) Relative motion between the coating and counter surface, b) Complete peeling off of the coating from the substrate, c) Relative motion between the coating in a texture and a counter surface, d) Fracture of the top portion of the coating in the textured surface.

Table 1. Summary of some of the LST process parameters and their values to enhance the adhesion behavior.

Author	Parameters	Values
B. He *et al.* [20]	Energy density	2.4 J cm^{-2}
	Shots per area	5 to 10 μm
	Pulse repetition rate	10 and 20 Hz

(Table 1) cont.....

T. Enomoto and T. Sugihara [23]	Peak wavelength	800 nm
	Pulse width	150 fs
	Cyclic frequency	1 kHz
	Pulse energy	300 μJ
L. Romoli *et al.* [25]	Laser power	12-18 W
	Tangential scan speed	100-1000 mm/s
	Hatch distance	0.035-0.200 mm
	Pulse overlap	20 kHz
	Energy density	0.343-6.912 J mm^{-2}
X. Zheng *et al.* [28]	Number of repeats of laser pulse processing	20, 40, 200
	Frequency	20, 30 kHz

EFFECT OF PARAMETER ON WETTABILITY

Hydrophobicity of the substrate is the attractive nature of the liquid layer. Different parameters influence the hydrophobicity or wettability of the substrate. This plays a vital role in determining the efficiency of the textured surface [29, 30]. P.W. Shum *et al.* [29, 31] used LST on DLC coated 316L stainless steel. The effect of textures on the hydrophobicity, non-stickiness, and wear resistance of DLC coated 316L stainless steel was investigated. Fig. (**7a**) shows this relationship between the dimple density and b/a (where 'b' is the width of the grooves and 'a' is the distance between two grooves) ratio with contact angle. Fig. (**7b**) reveals anincrease in dimple density percentage and the b/a ratio. The expansion of gas leads to temperature reduction, condensation of the moisture, and absorption on the gas lubricating surface that destroys the gas lubricating film. During vapor phase lubrication at high pressure and low temperature, the stable running of the gas seal is monitored in hydrophobic seals [32]. To improve the efficiency of hydrophobicity and increase the wetting behavior of the SiC surface, which is hydrophilic in nature, micro square convex laser texturing is done on the material surface. The increase in surface roughness increases the hydrophobic property of the material. The contact angle of the liquid also increases with an increase in surface roughness. Fig. (**8**) schematically explains the contact angle variation due to an increase in surface roughness. To avoid condensation of the water droplets on the SiC surface, it is good to texture the surface with a micro protuberant structure [33]. Chun *et al.* [34] carried out experiments to fabricate superhydrophobic metallic surfaces by two methods, namely nanoscale texturing and annealing. However, the conversion of hydrophilic to superhydrophobic was not that easy; instead of days, it took months to achieve this particular property as the surface was exposed to ambient conditions. The experiment was carried out over copper and brass substrate.

Superficial post-processing was introduced to reduce the wettability time.

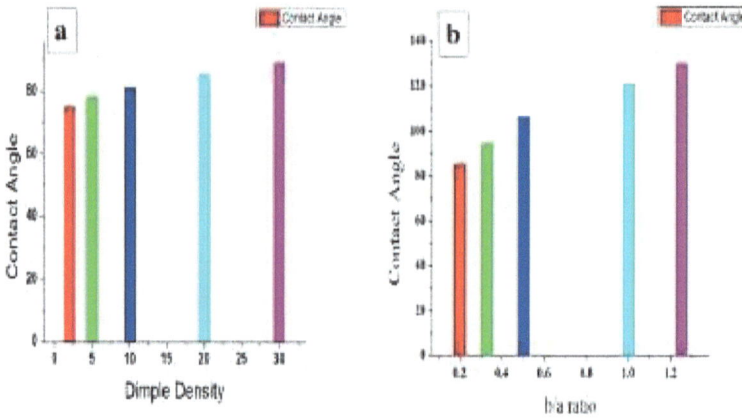

Fig. (7). Effect of dimple density and b/a ratio on contact angle variation.

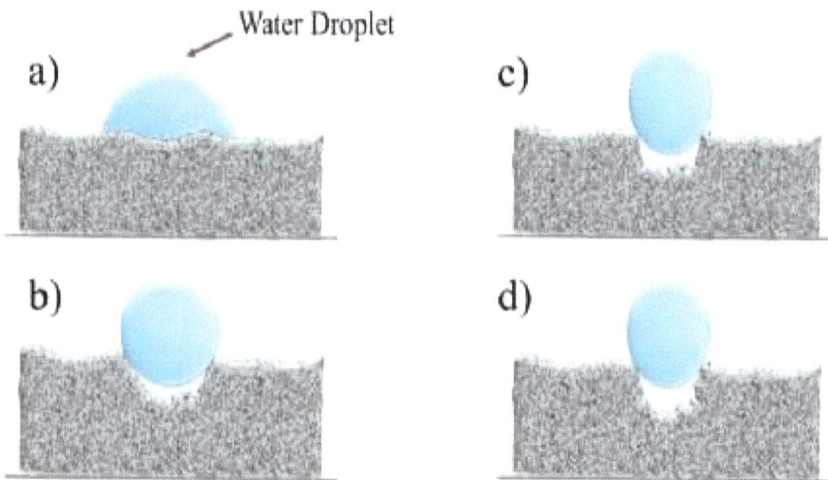

Fig. (8). Schematic representation of contact angle variation: **a)** Ra = 0.1 μm; θ=89.4º, **b)** Ra = 1 μm; θ = 102.8º, **c)** Ra = 2 μm; θ=109.1º, **d)** Ra = 3 μm; θ=119.1º.

This conversion of hydrophilic to superhydrophobic was carried out at a 100 °C low-temperature annealing process between copper and brass surface. Laser texturing is carried out over the zirconium surface with TiC and B$_4$C with a 3% contribution each. In this process, hydrophobicity is increased by the formation of surface texturing by micro and nano poles and relative cavities are formed by the laser melting process. The contact angle after laser melting with TiC and B$_4$C drastically improved to 130°, and subsequent microhardness and fracture toughness were also increased. The friction coefficient was also found to be more, and it increased the hydrophobicity of the textured surface [35].

V.D. Ta *et al.* [36] had fabricated a superhydrophobic texture of contact angle of ~154° and contact hysteresis of ~4° on the 304S15 stainless steel by nanosecond laser texturing system. LST experiments revealed that the surface morphologies are greatly affected by laser power. An increase in laser fluence increases the depth of the textures. In connection with that, surface roughness increases with the laser fluence. The surface roughness directly controls the contact angle of the textured surfaces. Higher surface roughness yields a higher contact angle and, in turn, increases the superhydrophobic nature. Exposing textures in the ambient atmosphere over time will alter the wettability. In this work, wettability increases with the exposing time up to 22 days. This change in wettability relates to a change in surface chemistry. The surface roughness of the textures is also influenced by scan line spacing. The scan line spacing is the gap between the consecutive laser passes. Higher surface roughness was reported in lower scan line spacing and improved wettability was observed.

EFFECT OF TEXTURES ON THE FRICTION AND WEAR BEHAVIOR OF METALLIC MATERIALS

Excessive friction and resulting wear losses are the greatest threat to the mechanical components in service. Minimizing wear losses is an everlasting challenge for researchers and engineers in this field. As a surface modification technique, LST shows a prominent role in controlling the friction and wear in metallic materials [37]. In this section, other major parameters that cause friction and wear in different aspects, such as micro and nanoscale, are discussed [38, 39]. Kim *et al.* discussed the effect of texturing over frictional behavior based on different texture geometry and micro dimple distribution for automotive applications. The most significant factor on the wear rate was found to be the aspect ratio of the dimples and the effect of surface density was found to be very less for the coefficient of friction [40]. The piston ring friction analysis was compared with textured and non-textured piston cylinders and it was found that laser surface textured ring has a 25% reduction in friction than normal cylinder without compromising the efficiency [41]. Texturing topology was found to control the friction and the laser beam was used to produce dimples on the surface. The laser texturing was performed both in air and underwater medium and the thermal effect was found to be suppressed in the underwater medium. The dimples texture created reduced the coefficient of friction and the wear rate of the material [42]. The microhardness of the steel was increased to reduce the friction and wear on the steel surface. This process was carried out in nitrate steel with predominant microstructural development due to the laser surfacing. This change in microhardness property is measured using an atomic force microscope, resulting in increased microhardness and decreased wear rate of the textured surface [43]. Cobalt-chromium is the primary material for bone implants; every year around billions of people are undergoing hip replacement surgery. The major problem arises after 15 years due to wear debris diseases. Therefore, it is highly essential to reduce the wear that leads to wear debris by increasing the hardness of the material. Apart from considering the material to be ultra-smooth, there is an option of increasing the load-bearing capacity by introducing texture. The micro-textured pattern on articulating with polyethylene acetabular liner reduces the friction and wear compared to conventional surrogate prosthetic hip joint implants [44]. The general relation between tactile friction and the surface texture was investigated and using the friction against the finger pad, the geometrical feature and spacing of texturing were determined. The coefficient of friction was also found to decrease with a consecutive increase in normal load. Scale-dependent elastic behavior mainly influences the relation between tactile friction and surface texture in the top layer of the material [45].

Fan *et al*. [46] discussed the Al_2O_3 surface, which was textured with 5 types of micro dimples, and Si–DLC film was used as acoating. It was found that friction and wear of laser texturing are effective in achieving composite lubricant. This helped in reducing the friction factor drastically from 0.51 to 0.06. Wang [47] characterized dimples of micro-corner-cube shape fabricated on medical needles. Insertion test determined the geometrical feature on tribological behavior and it was found that dimples increase friction between phantom tissue and needle due to stress concentration. Dimple spacing is directly proportional to drag increment and area density is controlled by the geometry of the dimple. Liu *et al*. [48] discussed the ring cutting and pulse processing in the fabrication of micro-texture with merits and demerits of the processes. The rise in laser pulse increases the crater depth and diameter and pulsating affects the crater depth. In the ring cutting processing method, the diameter of the micro-pit changes with scan radium, and the depth changes with the average power. Qiu *et al*. [49] conducted a series of experiments on circular and elliptical dimples with various sizes, densities, and depth to diameter ratios. A low coefficient of friction is experienced on the textured specimens compared to plain (dimple-free) surfaces. Sugihara *et al*. [50] reported an increase in tool life using texturing. The stripe-grooved surfaces formed on the tool's rake face were found to increase the tool life with higher wear resistance. The relation between texture dimensions and wear behavior was investigated. This was tested by a face milling experiment on steel materials, and the tool was found to reduce flank wear. The ratio of film thickness to surface roughness was found to decrease due to an increase in operating conditions. Mourier *et al*. [51] studied the effect of the single circular dimple on the EHL tribometer created by a femtosecond pulse laser. The experimental analysis of the EHL ball on the disk was performed by lubricating phenomena promoted by microvoids. The micro dimples that pass through the contact act as lubricant micro reservoirs, resulting in the modification of film thickness.

Q. Shang *et al.* [52] have fabricated dimple textures on the CuSn6 tin bronze specimens by LST. The wear behavior of the un-textured and textured specimens was investigated. The results had provided surprising information that the textured specimens showed higher wear losses than the un-textured specimens. The cause for this severe wear on textured surfaces was the materials softening of laser dimple edges. Shrinkages, porosities, cracks, and inverse segregation in the heat-affected zone were the reasons for the materials softening. This constraint was rectified by the work hardening process. This research work provides an insight into the properties of the material that are significant factors in determining the wear behavior of textured surfaces. The texture fabrication environment affects the texture quality. J. Hu and H. Xu [42] had used an LST in both air and underwater medium on the stainless steel surface. Underwater texturing significantly improved the quality of textures than those fabricated in a normal atmosphere. Water reduces the thermal effect of the laser and the recoil pressure of evaporated water increases the height/depth of texture dimensions. In the higher relative speed and normal load, the micro dimples acted as wear debris reservoirs and acted as surface lubricants. Y. Xing *et al.* [37] discussed the wear improvement mechanism of the texture in their research work of texturing experiments on Al_2O_3/TiC ceramic surface by LST. The waviness and linear grooved textures were fabricated. The coefficient of friction of the contact surface is represented in equation 2.

$$\mu = \mu_a + \mu_p + \mu_d \ (2)$$ **(2)**

Where,

μ = CoF of the contact surface

μ_a = CoF of adhesive friction

μ_p = CoF of plowing friction

μ_d = CoF induced by deformation and collision of asperities

Higher frictional coefficients with lower wear rates were observed in textured surfaces. Surface textures reduce the real contact area and increase the surface roughness by micro bulged asperities. These micro bulged asperities cold-welded with the counter surface asperities and offered resistance. Therefore, the frictional coefficients were increased in textured surfaces. On un-textured surfaces, the deformed asperities acted like a third body abrasive between the counter surface and substrate. These deformed asperities promoted the rolling friction, which reduced the frictional coefficient. However, in the case of textured surfaces, these deformed asperities got trapped into their cavities and avoided the rolling friction.

Therefore, the frictional coefficient increased by these trapped asperities. Fig. (**8**) schematically explains this phenomenon.

Additionally, the micro-cutting effect of the textures also increased the surface roughness and increased the frictional coefficient. Moreover, texture densities also play a significant role in the increase in frictional coefficient. An increase in texture densities per square area increases the surface roughness. This increased surface roughness imparts higher frictional forces for asperities deformation. Most of the shapes of the textures used for wear improvement would be in the form of dimples and grooves. Table **2** provides some insight into the forms of textures and their fabricated process parameters.

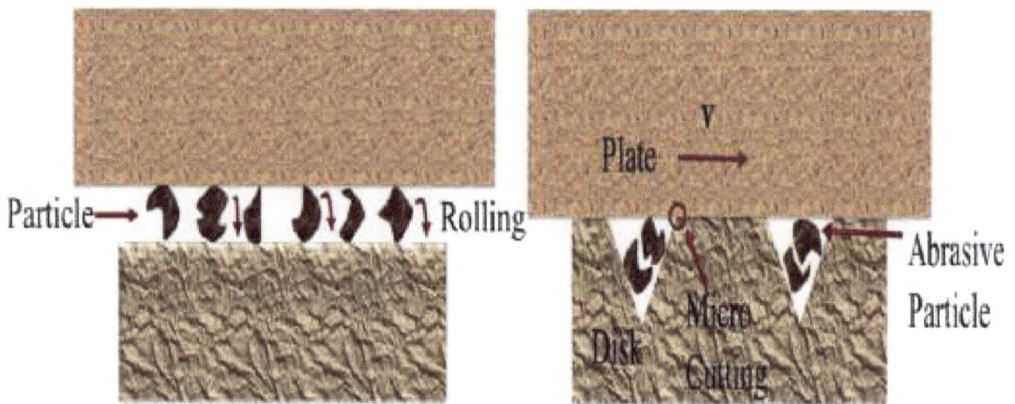

Fig. (9). Wear mechanism of **a)** Smooth surface, **b)** Textured surface.

Table 2. Forms of textures and their fabrication process parameters.

Texture type	Material	Parameters
Waviness and linear groove [37]	Al$_2$O$_3$/TiC ceramic	Laser power: 16.2 W Frequency: 6000 Hz Scanning speed: 5 mm/s
Transverse grooves, crosshatch, parallel grooves, [53]	AISI52100 steel	Pulse duration: 10 ps Frequency: 10 kHz Wavelength: 355 nm Power: 5 μJ
Parabolic [54]	Cold work steel (Powder metallurgy route)	Number of shots: ~100 Laser energy: 70 μJ
Rectangular dimples [55]	Cast iron	Wavelength: 1064 nm Pulse duration: 100 ns Fluence: 4.2 J/cm^2

(Table 2) cont.....

Circular dimple [52]	CuSn6 bronze	Electric current: 165A Pulse width: 1 ms Laser frequency: 2 Hz Scan speed: 2 mm/s
Dimple and groove [56]	Tungsten carbide	Power: 40 W Frequency: 8300 Hz Pulse width: 220 ns Scanning speed: 1 m/s

EFFECT OF TEXTURING ON LUBRICATION REGIMES

Lubrication plays an important role in controlling the friction and wear rate of the sample. The lubrication is of different types based on texture geometry, solid lubrication, and liquid lubrication, *etc.* Mostly liquid lubricant is used for low wear material when the friction and wear rate is maximum, and solid or geometric texturing is used to control the wear by friction. In the reciprocating wear, tester dimple texture created using LST was tested. The textured and non-textured surfaces with metal to metal and metal to plastic combinations were tested for efficiency. The lubricant introduced in both smooth and textured surfaces was found to have good results in the textured surface as the distance between the asperities increased due to unidirectional sliding of lubrication over the dimpled texture. This leads to a positive result of lubrication with a textured surface [57 - 59]. The creation of microspores over the water lubrication enhances the wear resistance as the spore of size 150 microns experiences the transferred form of hydrodynamic lubrication to mixed lubrication, thereby reducing the friction coefficient [60]. The solid lubricant in the form of diamond-like carbon coating is used to improve the wear resistance.

The dimples are introduced perpendicular to the surface and diamond-like carbon is coated over the textured surface using the sputtering technique. The dimple density and diameter are varied at a range of over 30% and 300 micrometers and tested using a reciprocating slide wear tester using oil as a lubricant. The best wear resistance output-friction co-efficient was found to at the range of 10% dimple density and 100 micrometers diameter. This process reduces friction rate to 20% and wear rate to 52% [61]. The tribological property of the steel is improved with mixed lubrication for the dimple range of around 800 microns under mixed lubrication conditions. The experiment was conducted in the unidirectional condition of sliding at around 100°C with a constant depth to diameter ratio and found around 80% reduction in friction. The friction rate during lubrication mainly depends on the temperature factor; on reducing the oil temperature from 100 to 50 °C, a drastic reduction in the friction rate was found [62]. The oil film life cycle over the laser patterned stainless steel was carried out and calculated for 1000 sliding cycles. The lifetime of the oil film increases with a decrease in thickness of the film, and the pattern structure of around 6 μm increases the sliding cycle factor to 130 than that of normal untextured film. The wear debris produced is topographically trapped in a minimal position, which reduces the abrasion of the fil due to sliding wear of debris, thus increasing the lifetime of the oil film [63]. The dimples are used for generating hydrodynamic pressure between oil lubricated and the contact surface; the impact of a textured surface on the lubricating regime was investigated. Two oil with different viscosity were used to study the effect of lubrication over abrasion or wear. Laser texturing is found to be effective during higher speed and load with a high viscosity oil medium [64]. The soft elastohydrodynamic lubrication was analyzed using spherical micro dimples. The viscous lubrication is used between the smooth elastomer and rigid lubrication. The equation of elasticity and Reynolds equation provides the solution for pressure distribution and elastic deformation of the elastomer. The analysis showed that soft elastohydrodynamic lubrication reduces friction with increased load capacity [65]. Analytical and experimental tests were conducted to enhance the hydrostatic and hydrodynamic lubrication effect of sliding wear control.

The disk with dimples having different depth and densities was evaluated with oil lubrication having various viscosity ranges. The disk consisting of higher dimple density was found to produce more abrasive wear on the ball specimen. The maximum wear rate leads to quick conformal contact and transits from boundary to mixed lubrication condition, resulting in reduced friction coefficient by increasing wear in the ball [66]. Starved lubrication condition is carried out both in laser texturing and laser peening to improve the wear resistance using different parametric conditions. In laser texturing, the different parametric conditions, such as feature depth, area fraction, diameter, sliding direction, and length, were evaluated using the DoE approach. The major influencing factors on the coefficient of friction are found to be diameter and area fraction. It is observed that the lifetime of the material is increased using micro-texture under starved lubrication conditions. The optimal parametric condition was found to be a diameter of 100 μm, an area fraction of 5%, a depth of 50 μm, and a length of 500 μm maintaining sliding direction perpendicular to micro textures [67]. When the laser peening method was applied under starved lubrication condition, it was found that frictional performance was close to laser texturing condition. The supportive characterization study using microstructural and EDS observations revealed that the starved lubrication reduces both abrasive and adhesive wear better than untextured conditions [68]. With this varying condition, it is noted to have different conditions of lubrication based on texturing, solid, liquid, and starved lubrication effect over the wear mechanism, which are discussed in detail.

MODELLING STUDIES ON LASER TEXTURING

A numerical investigation was carried out to understand the tribological performance of textured piston ring-liner using the 2D Reynold equation, which is solved with a mass conservation algorithm. The surface texturing-induced increase in oil transport to the combustion chamber remained minimal [69]. The hydrodynamic support for the cylinder bore was modeled with a one-dimensional model to simulate the effect of engine cylinder top, oil control, and bore. The simulation results revealed that micro dimples were capable of producing hydrodynamic support to decrease wear and friction. The micro dimples experienced a special benefit on the flat surface [70]. Another simulation work was carried out with a microgroove on the piston ring liner. The averaged Reynolds equation with Patir and Cheng's flow factors were used for the numerical approach and mass conservation cavitation model by Elrod Adams. The Greenwood and Tripp methods were used to determine the contact pressure that arises from the asperity interactions.

The simulation results showed an increase in friction when the pocket enters the contact and decreases while leaving the contact. The simulation also expels three mechanisms that are due to inlet gap, suction, and film thickness. These three mechanisms control the response of micro-texture under all lubrication regimes [71]. The modeling of the rolling interface found an increase in the reduction thickness from 5% to 10%; a new channel was formed in the saddle point due to pressurization of lubrication as the contraction of sheet surface resulted in the connection of the original valley and new valley. Further, a 13% tight reduction lubricant present in the initial valley played a moderate role in the contact, providing deformation and roughness [72]. A simulation process that models the indentation of machining for the betterment of medical appliances was carried out, considering the LST overlap factors using a second-order regression equation. Using the least square support vector, a predictive model was formulated for determining the machine depth. The optimal parameter determination was done by leave-one-out cross-validation. The machined geometry of the micro-channels is predicted by the LS-SVM model for medical needles [73].

SCOPE FOR THE FUTURE RESEARCH

LST can be employed to understand the newly emerging materials. Especially, LST has not yet been extensively studied on the tribological properties of magnesium and its alloys. Moreover, most of the texture forms used for tribological property for the enhancement of metallic materials were found to be dimples and grooves. The effect of micro pillar-like textures is not widely reported. The effect of micro pillar-like textures on the tribological properties of metallic materials can be investigated in the future research work.

CONCLUDING REMARKS

LST has found a prominent position among the surface modification techniques in recent years, especially for the enhancement of frictional and wear properties of metallic materials. Along with the wear resistance improvement, LST enhances the adhesion strength of the coating with the substrate and improves the wettability characteristics. The commonly used surface texture form for the wear resistance improvement, adhesion improvement, and conversion of surfaces to superhydrophobic are found to be dimples and grooves. Improvement in the coating adhesion could be achieved by filling and consolidating coating materials into the cavities of the textures. The consolidated coating resisted the shear failure and experienced a minimal fracture at its top layer instead of completely peeling off from the substrate. Surface roughness is the main factor in monitoring wettability. Wettability increases with an increase in surface roughness. LST significantly increases the surface roughness of the substrate. Therefore, the superhydrophobic surfaces could be possible by the LST. LST has been utilized as an efficient tool to improve the frictional and wear resistance of metallic materials. During the wear process, the wear debris is entrapped into the cavities of the texture dimples or grooves and restricted the rolling friction caused by them. Therefore, the frictional coefficients are higher in textured surfaces compared to untextured surfaces. These entrapped wear debris might reduce the real contact area and thus restrict the wear losses. Texture cavities hold the lubricant and allow them for effective lubrication. Researchers are now focusing on numerically simulating the effect of textures on different applications. This would facilitate the effective design of textures for targeted purposes. The future scope of the LST relies on the investigation of their effect on emerging materials like magnesium. Micro pillar-like textures can also be attempted to investigate their effect on the tribological properties of metallic materials.

CONSENT FOR PUBLICATION

Not applicable.

CONFLICT OF INTEREST

The author declares no conflict of interest, financial or otherwise.

ACKNOWLEDGEMENTS

Declared none.

REFERENCES

[1] Mao B, Siddaiah A, Liao Y, Menezes PL. Laser surface texturing and related techniques for enhancing tribological performance of engineering materials: A review. J Manuf Process 2020; 53: 153-73.
[http://dx.doi.org/10.1016/j.jmapro.2020.02.009]

[2] Martz LS. Preliminary report of developments in interrupted surface finishes. Proc- Inst Mech Eng 1949; 161(1): 1-9.
[http://dx.doi.org/10.1243/PIME_PROC_1949_161_007_02]

[3] Anno JN, Walowit JA, Allen CM. Microasperity lubrication. 1968: 351-355.
[http://dx.doi.org/10.1115/1.3601568]

[4] Priest M, Taylor CM. Automobile engine tribology—approaching the surface. Wear 2000; 241(2): 193-203.
[http://dx.doi.org/10.1016/S0043-1648(00)00375-6]

[5] Holmberg K, Mathews A. Coatings tribology: a concept, critical aspects and future directions. Thin Solid Films 1994; 253(1-2): 173-8.
[http://dx.doi.org/10.1016/0040-6090(94)90315-8]

[6] Menezes PL, Kailas SV. Influence of surface texture on coefficient of friction and transfer layer formation during sliding of pure magnesium pin on 080 M40 (EN8) steel plate. Wear 2006; 261(5-6): 578-91.
[http://dx.doi.org/10.1016/j.wear.2006.01.001]

[7] He Y, Zou P, Zhu Z, *et al.* Design and application of a flexure-based oscillation mechanism for surface texturing. J Manuf Process 2018; 32: 298-306.
[http://dx.doi.org/10.1016/j.jmapro.2018.02.017]

[8] Prasad KN, Syed I, Subbu SK. Laser dimple texturing–applications, process, challenges, and recent developments: a review. Aust J Mech Eng 2019; 29: 1-6.
[http://dx.doi.org/10.1080/14484846.2019.1705533]

[9] Singh A, Harimkar SP. Laser surface engineering of magnesium alloys: a review. JOM 2012; 64(6): 716-33.
[http://dx.doi.org/10.1007/s11837-012-0340-2]

[10] Kennedy E, Byrne G, Collins DN. A review of the use of high power diode lasers in surface hardening. J Mater Process Technol 2004; 155: 1855-60.
[http://dx.doi.org/10.1016/j.jmatprotec.2004.04.276]

[11] Luo KY, Yao HX, Dai FZ, Lu JZ. Surface textural features and its formation process of AISI 304 stainless steel subjected to massive LSP impacts. Opt Lasers Eng 2014; 55: 136-42.
[http://dx.doi.org/10.1016/j.optlaseng.2013.10.026]

[12] Jiaa CL. Spectrogram analysis of random laser texture pattern media. Surf Coat Tech 2000; 123(2-3): 140-6.
[http://dx.doi.org/10.1016/S0257-8972(99)00476-4]

[13] Samanta A, Wang Q, Shaw SK, Ding H. Roles of chemistry modification for laser textured metal alloys to achieve extreme surface wetting behaviors. Mater Des 2020; 30: 108744.
[http://dx.doi.org/10.1016/j.matdes.2020.108744]

[14] Xue X, Lu L, Wang Z, Li Y, Guan Y. Improving tribological behavior of laser textured Ti-20Zr-10-b-4Ta alloy with dimple surface. Mater Lett 2021; 305(15): 130876.
[http://dx.doi.org/10.1016/j.matlet.2021.130876]

[15] Etsion I. State of the art in laser surface texturing. J Tribol 2005; 127(1): 248-53.
[http://dx.doi.org/10.1115/1.1828070]

[16] Zhou J, Shen H, Pan Y, Ding X. Experimental study on laser microstructures using long pulse. Opt Lasers Eng 2016; 78: 113-20.

[http://dx.doi.org/10.1016/j.optlaseng.2015.10.009]

[17]   Dehghanpour HR, Parvin P, Abdolahi S. Performance enhancement of solar panel by surface texturing using ArF excimer laser. Optik (Stuttg) 2015; 126(24): 5496-8.
[http://dx.doi.org/10.1016/j.ijleo.2015.09.108]

[18]   Etsion I, Sher E. Improving fuel efficiency with laser surface textured piston rings. Tribol Int 2009; 42(4): 542-7.
[http://dx.doi.org/10.1016/j.triboint.2008.02.015]

[19]   Mangano C, De Rosa A, Desiderio V, *et al.* The osteoblastic differentiation of dental pulp stem cells and bone formation on different titanium surface textures. Biomaterials 2010; 31(13): 3543-51.
[http://dx.doi.org/10.1016/j.biomaterials.2010.01.056] [PMID: 20122719]

[20]   He B, Petzing J, Webb P, Leach R. Improving copper plating adhesion on glass using laser machining techniques and areal surface texture parameters. Opt Lasers Eng 2015; 75: 39-47.
[http://dx.doi.org/10.1016/j.optlaseng.2015.06.004]

[21]   Sugihara T, Enomoto T. Improving anti-adhesion in aluminum alloy cutting by micro stripe texture. Precis Eng 2012; 36(2): 229-37.
[http://dx.doi.org/10.1016/j.precisioneng.2011.10.002]

[22]   Sugihara T, Enomoto T. Development of a cutting tool with a nano/micro-textured surface—Improvement of anti-adhesive effect by considering the texture patterns. Precis Eng 2009; 33(4): 425-9.
[http://dx.doi.org/10.1016/j.precisioneng.2008.11.004]

[23]   Enomoto T, Sugihara T. Improvement of anti-adhesive properties of cutting tool by nano/micro textures and its mechanism. Procedia Eng 2011; 19: 100-5.
[http://dx.doi.org/10.1016/j.proeng.2011.11.086]

[24]   Oka M. Stiction problem of annular-shaped laser textured bump on a hard disk. Tribol Int 2000; 33(5-6): 353-6.
[http://dx.doi.org/10.1016/S0301-679X(00)00052-9]

[25]   Aurich JC, Bohley M, Reichenbach IG, Kirsch B. Surface quality in micro milling: Influences of spindle and cutting parameters. CIRP Ann 2017; 66(1): 101-4.
[http://dx.doi.org/10.1016/j.cirp.2017.04.029]

[26]   Zhao Q, Talke FE. Effect of environmental conditions on the stiction behavior of laser textured hard disk media. Tribol Int 2000; 33(3-4): 281-7.
[http://dx.doi.org/10.1016/S0301-679X(00)00043-8]

[27]   Kromer R, Costil S, Verdy C, Gojon S, Liao H. Laser surface texturing to enhance adhesion bond strength of spray coatings–cold spraying, wire-arc spraying, and atmospheric plasma spraying. Surf Coat Tech 2018; 352: 642-53.
[http://dx.doi.org/10.1016/j.surfcoat.2017.05.007]

[28]   He ZH, Ma ZG, Luo YY, *et al.* An experimental investigation of the off-diagonal Seebeck effect on textured YBCO high-Tc superconductor. Appl Supercond 1998; 4(12): 619-23.
[http://dx.doi.org/10.1016/S0964-1807(97)00025-2]

[29]   Shum PW, Zhou ZF, Li KY. To increase the hydrophobicity, non-stickiness and wear resistance of DLC surface by surface texturing using a laser ablation process. Tribol Int 2014; 78: 1-6.
[http://dx.doi.org/10.1016/j.triboint.2014.04.026]

[30]   Razi S, Madanipour K, Mollabashi M. Laser surface texturing of 316L stainless steel in air and water: A method for increasing hydrophilicity *via* direct creation of microstructures. Opt Laser Technol 2016; 80: 237-46.
[http://dx.doi.org/10.1016/j.optlastec.2015.12.022]

[31]   Prakash FP, Jeyaprakash N, Duraiselvam M, Prabu G, Yang CH. Droplet spreading and wettability of laser textured C-263 based nickel superalloy. Surf Coat Tech 2020; 397: 126055.

[http://dx.doi.org/10.1016/j.surfcoat.2020.126055]

[32] Hu YZ, Ma TB. Tribology of nanostructured surfaces. Comprehensive Nanoscience and Nanotechnology 2019; 1–5: pp. 309-42.

[33] Ma C, Bai S, Peng X, Meng Y. Improving hydrophobicity of laser textured SiC surface with micro-square convexes. Appl Surf Sci 2013; 266: 51-6.
      [http://dx.doi.org/10.1016/j.apsusc.2012.11.068]

[34] Chun DM, Ngo CV, Lee KM. Fast fabrication of superhydrophobic metallic surface using nanosecond laser texturing and low-temperature annealing. CIRP Ann 2016; 65(1): 519-22.
      [http://dx.doi.org/10.1016/j.cirp.2016.04.019]

[35] Yilbas BS. Laser texturing of zirconia surface with presence of TiC and B4C: Surface hydrophobicity, metallurgical, and mechanical characteristics. Ceram Int 2014; 40(10): 16159-67.
      [http://dx.doi.org/10.1016/j.ceramint.2014.07.047]

[36] Jamshidi YT, Anaraki AP, Sadighi M, Kadkhodapour J, Mirbagheri SM, Akhavan B. Micro-structure analysis of quasi-static crushing and low-velocity impact behavior of graded composite metallic foam filled tube. Met Mater Int 2019; 4: 1-4.
      [http://dx.doi.org/10.1007/s12540-019-00502-0]

[37] Xing Y, Deng J, Feng X, Yu S. Effect of laser surface texturing on Si3N4/TiC ceramic sliding against steel under dry friction. Materials & Design (1980-2015) 2013; 52: 234-45.

[38] Dunn A, Carstensen JV, Wlodarczyk KL, *et al.* Nanosecond laser texturing for high friction applications. Opt Lasers Eng 2014; 62: 9-16.
      [http://dx.doi.org/10.1016/j.optlaseng.2014.05.003]

[39] Xing Y, Deng J, Wu Z, Wu F. High friction and low wear properties of laser-textured ceramic surface under dry friction. Opt Laser Technol 2017; 93: 24-32.
      [http://dx.doi.org/10.1016/j.optlastec.2017.01.032]

[40] Kim B, Chae YH, Choi HS. Effects of surface texturing on the frictional behavior of cast iron surfaces. Tribol Int 2014; 70: 128-35.
      [http://dx.doi.org/10.1016/j.triboint.2013.10.006]

[41] Ryk G, Etsion I. Testing piston rings with partial laser surface texturing for friction reduction. Wear 2006; 261(7-8): 792-6.
      [http://dx.doi.org/10.1016/j.wear.2006.01.031]

[42] Hu J, Xu H. Friction and wear behavior analysis of the stainless steel surface fabricated by laser texturing underwater. Tribol Int 2016; 102: 371-7.
      [http://dx.doi.org/10.1016/j.triboint.2016.06.001]

[43] Gualtieri E, Borghi A, Calabri L, Pugno N, Valeri S. Increasing nanohardness and reducing friction of nitride steel by laser surface texturing. Tribol Int 2009; 42(5): 699-705.
      [http://dx.doi.org/10.1016/j.triboint.2008.09.008]

[44] Chyr A, Qiu M, Speltz J, Jacobsen RL, Sanders AP, Raeymaekers B. A patterned microtexture to reduce friction and increase longevity of prosthetic hip joints. Wear 2014; 315(1-2): 51-7.
      [http://dx.doi.org/10.1016/j.wear.2014.04.001] [PMID: 25013240]

[45] van Kuilenburg J, Masen MA, Groenendijk MN, Bana VV, van der Heide E. An experimental study on the relation between surface texture and tactile friction. Tribol Int 2012; 48: 15-21.
      [http://dx.doi.org/10.1016/j.triboint.2011.06.003]

[46] Fan H, Zhang Y, Hu T, Song J, Ding Q, Hu L. Surface composition–lubrication design of Al2O3/Ni laminated composites—Part I: Tribological synergy effect of micro–dimpled texture and diamond–like carbon films in a water environment. Tribol Int 2015; 84: 142-51.
      [http://dx.doi.org/10.1016/j.triboint.2014.12.016]

[47] Wang X, Giovannini M, Xing Y, Kang M, Ehmann K. Fabrication and tribological behaviors of

corner-cube-like dimple arrays produced by laser surface texturing on medical needles. Tribol Int 2015; 92: 553-8.
[http://dx.doi.org/10.1016/j.triboint.2015.07.042]

[48]   Li D, Chen X, Guo C, *et al.* Micro surface texturing of alumina ceramic with nanosecond laser. Procedia Eng 2017; 174: 370-6.
[http://dx.doi.org/10.1016/j.proeng.2017.01.155]

[49]   Qiu Y, Khonsari MM. Experimental investigation of tribological performance of laser textured stainless steel rings. Tribol Int 2011; 44(5): 635-44.
[http://dx.doi.org/10.1016/j.triboint.2011.01.003]

[50]   Sugihara T, Enomoto T. Crater and flank wear resistance of cutting tools having micro textured surfaces. Precis Eng 2013; 37(4): 888-96.
[http://dx.doi.org/10.1016/j.precisioneng.2013.05.007]

[51]   Mourier L, Mazuyer D, Lubrecht AA, Donnet C, Audouard E. Action of a femtosecond laser generated micro-cavity passing through a circular EHL contact. Wear 2008; 264(5-6): 450-6.
[http://dx.doi.org/10.1016/j.wear.2006.08.037]

[52]   Shang Q, Yu A, Wu J, Shi C, Niu W. Influence of heat affected zone on tribological properties of CuSn6 bronze laser dimple textured surface. Tribol Int 2017; 105: 158-65.
[http://dx.doi.org/10.1016/j.triboint.2016.10.008]

[53]   Vlădescu SC, Ciniero A, Tufail K, Gangopadhyay A, Reddyhoff T. Looking into a laser textured piston ring-liner contact. Tribol Int 2017; 115: 140-53.
[http://dx.doi.org/10.1016/j.triboint.2017.04.051]

[54]   Vilhena LM, Sedlaček M, Podgornik B, Vižintin J, Babnik A, Možina J. Surface texturing by pulsed Nd: YAG laser. Tribol Int 2009; 42(10): 1496-504.
[http://dx.doi.org/10.1016/j.triboint.2009.06.003]

[55]   Saeidi F, Parlinska-Wojtan M, Hoffmann P, Wasmer K. Effects of laser surface texturing on the wear and failure mechanism of grey cast iron reciprocating against steel under starved lubrication conditions. Wear 2017; 386: 29-38.
[http://dx.doi.org/10.1016/j.wear.2017.05.015]

[56]   Bhaduri D, Batal A, Dimov SS, *et al.* On design and tribological behaviour of laser textured surfaces. Procedia CIRP 2017; 60: 20-5.
[http://dx.doi.org/10.1016/j.procir.2017.02.050]

[57]   Jones K, Schmid SR. Experimental investigation of laser texturing and its effect on friction and lubrication. Procedia Manuf 2016; 5: 568-77.
[http://dx.doi.org/10.1016/j.promfg.2016.08.047]

[58]   Chang TL, Tsai TK, Yang HP, Huang JZ. Effect of ultra-fast laser texturing on surface wettability of microfluidic channels. Microelectron Eng 2012; 98: 684-8.
[http://dx.doi.org/10.1016/j.mee.2012.05.057]

[59]   Sudeep U, Tandon N, Pandey RK. Friction and vibration behaviors of lubricated laser textured point contacts under reciprocating rolling motion with highlights on the used laser parameters. Procedia Technology 2014; 14: 4-11.
[http://dx.doi.org/10.1016/j.protcy.2014.08.002]

[60]   Wang X, Kato K, Adachi K, Aizawa K. The effect of laser texturing of SiC surface on the critical load for the transition of water lubrication mode from hydrodynamic to mixed. Tribol Int 2001; 34(10): 703-11.
[http://dx.doi.org/10.1016/S0301-679X(01)00063-9]

[61]   Shum PW, Zhou ZF, Li KY. Investigation of the tribological properties of the different textured DLC coatings under reciprocating lubricated conditions. Tribol Int 2013; 65: 259-64.
[http://dx.doi.org/10.1016/j.triboint.2013.01.012]

[62]   Braun D, Greiner C, Schneider J, Gumbsch P. Efficiency of laser surface texturing in the reduction of friction under mixed lubrication. Tribol Int 2014; 77: 142-7.
[http://dx.doi.org/10.1016/j.triboint.2014.04.012]

[63]   Rosenkranz A, Heib T, Gachot C, Mücklich F. Oil film lifetime and wear particle analysis of laser-patterned stainless steel surfaces. Wear 2015; 334: 1-2.
[http://dx.doi.org/10.1016/j.wear.2015.04.006]

[64]   Kovalchenko A, Ajayi O, Erdemir A, Fenske G, Etsion I. The effect of laser surface texturing on transitions in lubrication regimes during unidirectional sliding contact. Tribol Int 2005; 38(3): 219-25.
[http://dx.doi.org/10.1016/j.triboint.2004.08.004]

[65]   Shinkarenko A, Kligerman Y, Etsion I. The effect of surface texturing in soft elasto-hydrodynamic lubrication. Tribol Int 2009; 42(2): 284-92.
[http://dx.doi.org/10.1016/j.triboint.2008.06.008]

[66]   Kovalchenko A, Ajayi O, Erdemir A, Fenske G. Friction and wear behavior of laser textured surface under lubricated initial point contact. Wear 2011; 271(9-10): 1719-25.
[http://dx.doi.org/10.1016/j.wear.2010.12.049]

[67]   Saeidi F, Meylan B, Hoffmann P, Wasmer K. Effect of surface texturing on cast iron reciprocating against steel under starved lubrication conditions: A parametric study. Wear 2016; 348: 17-26.
[http://dx.doi.org/10.1016/j.wear.2015.10.020]

[68]   Li K, Yao Z, Hu Y, Gu W. Friction and wear performance of laser peen textured surface under starved lubrication. Tribol Int 2014; 77: 97-105.
[http://dx.doi.org/10.1016/j.triboint.2014.04.017]

[69]   Usman A, Park CW. Numerical investigation of tribological performance in mixed lubrication of textured piston ring-liner conjunction with a non-circular cylinder bore. Tribol Int 2017; 105: 148-57.
[http://dx.doi.org/10.1016/j.triboint.2016.09.043]

[70]   Tomanik E, Profito FJ, Zachariadis DC. Modelling the hydrodynamic support of cylinder bore and piston rings with laser textured surfaces. Tribol Int 2013; 59: 90-6.
[http://dx.doi.org/10.1016/j.triboint.2012.01.016]

[71]   Profito FJ, Vlădescu SC, Reddyhoff T, Dini D. Transient experimental and modelling studies of laser-textured micro-grooved surfaces with a focus on piston-ring cylinder liner contacts. Tribol Int 2017; 113: 125-36.
[http://dx.doi.org/10.1016/j.triboint.2016.12.003]

[72]   Ike H, Tsuji K, Takase M. *In situ* observation of a rolling interface and modeling of the surface texturing of rolled sheets. Wear 2002; 252(1-2): 48-62.
[http://dx.doi.org/10.1016/S0043-1648(01)00854-7]

[73]   Wang X, Han P, Giovannini M, Ehmann K. Modeling of machined depth in laser surface texturing of medical needles. Precis Eng 2017; 47: 10-8.
[http://dx.doi.org/10.1016/j.precisioneng.2016.06.012]

CHAPTER 10

Material and Tribological Characterization of Surfaces

Kaushik N. Ch [1,*], **Krishna Kishore Mugada** [2,*] and **Muralimohan Cheepu** [3]

[1] *Department of Mechanical Engineering, School of Engineering and Technology, BML Munjal University, Gurgaon 122413, Haryana, India*

[2] *Department of Mechanical Engineering, Indian Institute of Technology Delhi, New Delhi 110016, India*

[3] *Super-TIG Welding Co., Limited, Busan - 46722, Republic of Korea*

Abstract: Surface, the top most or outermost layer of a given material, is very much influential in the initial interaction with other bodies or their surfaces. In tribological applications, surface interaction is important, therefore, it needs to be treated or protected for the improvement of properties. The surface treatment techniques are significantly important as they enhance various properties of the material, such as surface strength, surface hardness, surface roughness, friction and wear resistance, chemical resistance, and corrosion resistance, *etc*. These surface properties are studied with a wide range of characterization methods, such as the chemical composition of the surface being analyzed with the energy dispersive, Auger electron, glow discharge, optical emission spectroscopy and X-ray spectroscopy. The microstructure and morphology of the surface are studied by using light optical microscopy, scanning electron microscopy and transmission electron microscopy. In the present context, the tribological characterization includes the evaluation of surface roughness, friction and wear aspects of the surfaces from macro to nanoscale. The surface profile can be assessed by using contact and non-contact type surface profilometers. The friction aspects can be studied using a simple scratch tester or a multi-scale tribometer or by measuring lateral forces in atomic force microscopy. This chapter covers the theoretical aspects of various surface characterization techniques and tribological characterization methods.

Keywords: Atomic force microscopy, Friction and wear, Microstructure, Morphology, Optical, Profilometer, Scanning electron and transmission electron microscopy, Spectroscopy, Surface, Tribological characterization.

* **Correspondence author Kaushik N. Ch :** Department of Mechanical Engineering, School of Engineering and Technology, BML Munjal University, Gurgaon 122413, Haryana, India; Tel: +91-9030879876, Fax: 014010675; E-mail: kaushiknch1234@gmail.com, and **Krishna Kishore Mugada:** Department of Mechanical Engineering, Indian Institute of Technology Delhi, New Delhi 110016, India; Tel:+91-9505674467; Fax: 014010675; E-mail:mugada.krishnakishore@gmail.com

Jeyaprakash Natarajan and Che-Hua Yang (Eds.)
All rights reserved-© 2021 Bentham Science Publishers

INTRODUCTION

The surface of a material is the interface between the bulk and the external phase (solid, liquid, or gas) in direct contact with the material. The material surface can be understood in several ways, but in mechanical aspects, it is discussed depending on the type of interaction and its response after the interaction. The type of atoms present on the topmost or outermost layer of the material, formation of layers, depth of layer/layers formed when compared to bulk sample, and their physical and chemical properties during interaction has to be understood well to understand the performance characteristics. Erosion of surface is sometimes influenced by preferential surface segregation of species and processes occurring at the surface, such as oxidation and degradation [1].

In this chapter, the fundamental discussion related to metallurgical and/or microstructural characterization was done at first, and later, the tribological characterization was done theoretically. The microstructural characterization is essential to probe and map the surface and sub-surface structure characteristics of a material basically from the micro to nanoscale range. Understanding, controlling and characterizing the behavior of surfaces is the central theme of tribology. Friction and wear processes occur at surfaces and interfaces and are a manifestation of the physical and chemical characteristics of the materials in question. Surface modification is one of the means by which these properties and processes are controlled or mitigated. For example, the inherent wear properties of a material depend on parameters, such as hardness, the structure of the surface, and chemical reactivity. Wear behavior may be controlled by the application of hard surface films (thin or thick) and also by appropriate lubrication regimes. Analytical tribology necessarily involves the determination of the chemical, electronic, and structural characteristics of the surface and wear debris. Modern surface analytical techniques provide a comprehensive understanding of tribological mechanisms via spectroscopy, imaging, and depth profiling.

CHARACTERIZATION TECHNIQUES OVERVIEW

The selection of appropriate physical and chemical surface characterization of analyzing the material/specimen is important. No one analytical technique gives all the information required from the specimen. Therefore, a systematic approach is followed to select the analytical technique, and subsequent methods are adopted to characterize the sample. Fig. (**1**) shows the overview of the characteristics of materials (physical and chemical). The characterization strategy has the following stages:

- Identification and definition of the problem to be studied.
- Development of the hypotheses to the problem.
- Nature of information to be obtained from the characterization system.
- Usage of the available techniques to address the information required from the specimen.
- Confirmation of the reproducibility of the results and consistency of the findings from different techniques.

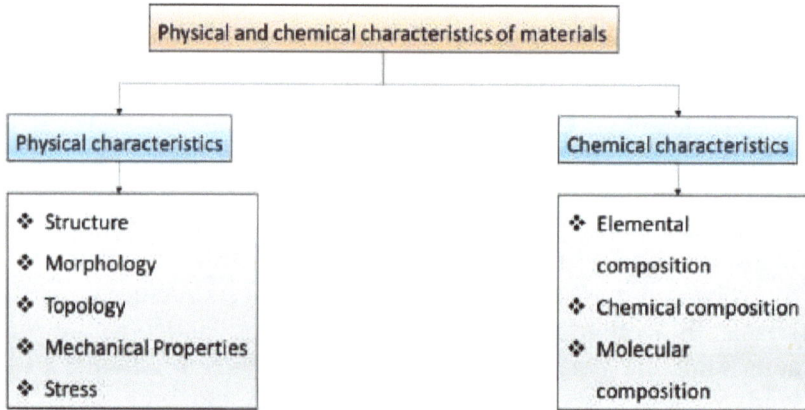

Physical and chemical characteristics of materials

Physical characteristics

- Structure
- Morphology
- Topology
- Mechanical Properties
- Stress

Chemical characteristics

- Elemental composition
- Chemical composition
- Molecular composition

Fig. (1). Physical and chemical characteristics of materials.

MATERIAL CHARACTERIZATION TECHNIQUES

Microstructural examinations of metals and alloys have been carried out for over a century. Microscopy is used to identify the details in the objects. Microscopy is a field that uses microscopes to view objects and details, which cannot be seen with the naked eye. Optical, electron, and scanning probe microscopy are the most prominent microscopy advancements.

Henry developed methods to view the samples under the reflected light microscope. The works by Henry Clifford Sorby have laid the foundation for the development of the light (optical) microscope. Optical/electron microscopy involves the diffraction, reflection/refraction of electromagnetic radiation/electron beams that interact with the object/specimen to view the details of the specimen. With the developments in understanding the material strength, the requirement of high-resolution microscopes to a level of inter-atomic spacing was found necessary. This led to the development of electron microscopy (Scanning electron and Transmission electron microscopy) [2 - 7]. Table **1** shows the history and developments in microscopy for materials characterization.

Table 1. History and development of microscopy.

Year	Development
1670	Antoni van Leeuwenhoek, Single-lens optical instruments developments, Father of microscopy
1800	J.J. Lister, Compound optical instruments development, and lens corrected for chromatic and spherical aberrations
1839	L. Daguerres invented and developed the silver plate method of the light microscope
1850	L Seidels, Theory of lens aberrations development
1858	Plucker observed the deflection of cathode rays
1864	H C Sorby, Microscopy observation in steels
1873	Ernst Abbe, Light wavelength limitations to resolve details in an object
1924	L. de Broglie, material wave prediction (theoretical)
1926	H Busch, Electromagnetic lens development
1927	C J Davisson, L H Germer- Electron diffraction discovery
1928	H Bethe, Diffraction theory
1932	Max Knoll and Ernst Ruska obtained images from the electron microscope prototype
1937	Manfred von Ardenne, Scanning electron microscope development
1939	Siemens produced the first electron microscope
1939	Max Knoll and Ernst Ruska, first commercial TEM developed
1956	W Bollmann and P B Hirsch observed dislocations in metal crystals
1960	P B Hirsch, Image contrast theory

This section briefly explains the light and electron microscopes for material characterization, as shown in Fig. (**2**). The light (optical) microscope uses visible light and a system of lenses to magnify the details of the smaller objects/specimens. Typically, a camera is attached to capture the image, which is also called a micrograph. The source of illumination in an electron microscope is a beam of accelerated electrons. As the wavelength of electrons is much higher compared to light photons, electron microscopes have a higher resolution compared to light microscopes. In electron microscopy, for instance, in the scanning electron microscopy, the secondary electrons emitted during the interaction of the electron beam with the specimen are detected by the secondary electron detector (Everhart Thornley detector) to view the micrograph of the specimen. The electron beam is scanned in a raster scan pattern, and the intensity of the detected signal and the position of the beam are combined to get the complete micrograph of the sample.

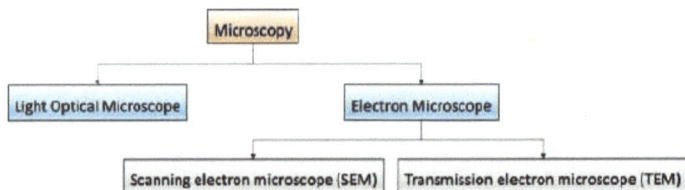

Fig. (2). Material characterization techniques.

Optical Microscopy

Magnifying objects using the glass lens started in the 13th century AD. In the year 1670, the development of the single-lens instruments began, and the Dutch microscopist Antoni van Leeuwenhoek pioneered these developments. He is also called the father of microscopy. Optical microscopy is defined as the formation of the magnified image of the selected location of an object or specimen using visible light as the illumination source and arrangement of the lens. It is widely used by researchers because it is cost-effective and easy to operate. The objective lens forms a magnified image of the selected region in the specimen, which is further magnified by the eyepiece lens.

The total magnification is the product of the magnifications of both objective and eyepiece lenses. For example, if the magnification of the objective lens is 100x and the eyepiece is 10x, the total magnification is 1000x. The disadvantage of the optical microscope is the resolution limit (min size is 0.2 microns). The limitation of the wavelength of visible light is around 0.5 microns.

The basic optical microscopes are a very simple and complex combination of the lens, which can be used to improve contrast and resolution. In general, the object is placed on a stage and viewed directly from one/two eyepieces in the microscope. A camera is attached to capture the image, which is also called a micrograph. A range of objective lens is usually placed on the turret and rotated to the required lens and the height is adjusted. There are two fundamental types of optical microscopes; (a) the simple microscope and (b) the compound microscope. Fig. (**3**) shows the setup of the simple microscope, where the optical power is obtained from the single lens.

However, as shown in Fig. (**4**), the compound microscope uses an additional objective lens near the specimen to collect the light signal. The objective lens focuses on the real image of the specimen inside the microscope (Image 1), and the image is then magnified with the eyepiece lens (Image 2) to get the enlarged image. As discussed, the objective lens forms the real image of the specimen, which is very important in the optical microscope. Some of the properties of the objective lens are discussed below.

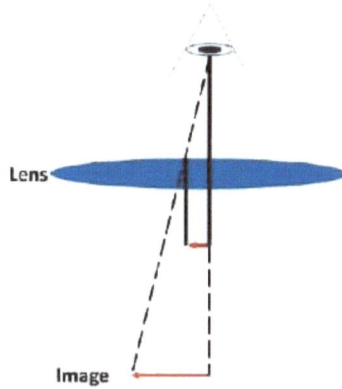

Fig. (3). Schematic representation of the simple microscope.

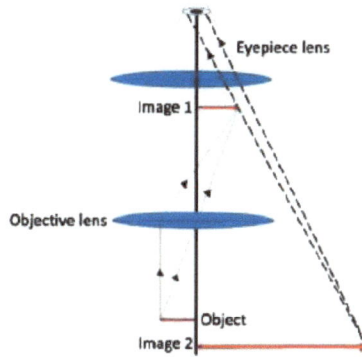

Fig. (4). Schematic representation of the compound microscope.

▪ Numerical Aperture (NA): It is a measure of the light collective function and is proportional to μ sin α.

NA = μ sin α

Where, is the refractive index of the medium between the specimen and the objective lens. is the semi angle of the most oblique rays entering from the front of the objective lens.

▪ Resolution: It is a function of numerical aperture, wavelength, and coherency of the light.

Abbe relation for the maximum resolution is given as:

Limit of resolution = $0.5\lambda / NA$ (for coherent illumination)

Resolving power = $2 \times NA / \lambda$

• Magnification: It is the ratio of numerical aperture on the objective side to the numerical aperture on the image side. The total magnification of the compound microscope is the product of the magnification of the objective lens and eyepiece lens. The minimum magnification required to resolve the images of the object is given by the ratio of the minimum distance between particles resolved by the eye to the minimum distance between particles resolved by the microscope. For example, if the power of the lens is 10 and the numerical aperture is 0.25, then it is written as 10x/0.25.

• Image brightness: It is dependent on the aperture on the object side of the lens system. In general, it is proportional to the square of the numerical aperture on the image space.

• Depth of field: It is the distance along the optical axis over which details of the specimen can be observed with adequate sharpness.

In general, the image contrast in an optical micrograph is improved by developing the differential local coefficient of scattering and reflection by polishing and *etc*hing. The optical methods for enhancing the contrast applicable to any kind of surface condition are given below:

• Bright field illumination

• Oblique illumination

• Darkfield illumination

• Polarized illumination

• Phase contrast illumination

• Multiple beam interferometry

• Interference contrast illumination

In summary, optical microscopes are widely used in various fields of science, from metallurgy to microbiology, medical and biotechnology, *etc*.

Scanning Electron Microscopy

When a beam of accelerated electrons is focused on the surface of the sample, the electrons interact with the atoms in the sample and the scattering and diffraction of the beam occur. The atoms excited by the electron beam are detected with various detectors to view   the   details of the specimen. Figs. (**5**) and (**6**)   shows

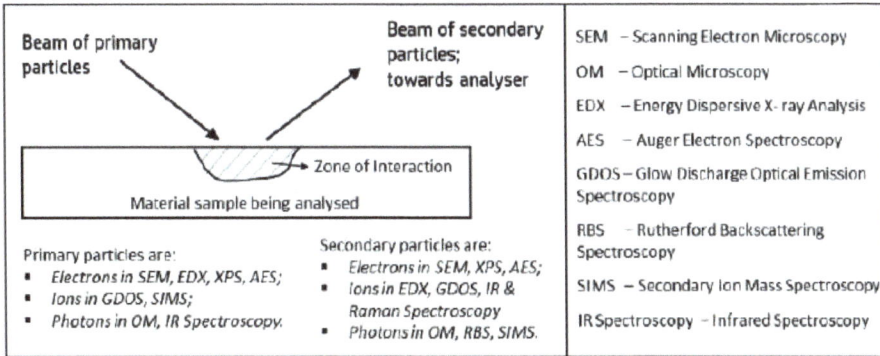

Beam of primary particles ↘	Beam of secondary particles; towards analyser ↗	SEM – Scanning Electron Microscopy
		OM – Optical Microscopy
	→ Zone of Interaction	EDX – Energy Dispersive X- ray Analysis
Material sample being analysed		AES – Auger Electron Spectroscopy
		GDOS – Glow Discharge Optical Emission Spectroscopy
Primary particles are:	Secondary particles are:	RBS – Rutherford Backscattering Spectroscopy
• *Electrons in SEM, EDX, XPS, AES;*	• *Electrons in SEM, XPS, AES;*	SIMS – Secondary Ion Mass Spectroscopy
• *Ions in GDOS, SIMS;*	• *Ions in EDX, GDOS, IR & Raman Spectroscopy*	IR Spectroscopy – Infrared Spectroscopy
• *Photons in OM, IR Spectroscopy.*	• *Photons in OM, RBS, SIMS.*	

Fig. (5). Schematic representation of the interaction of the beam of electrons and various ranges of signals in the bulk specimen.

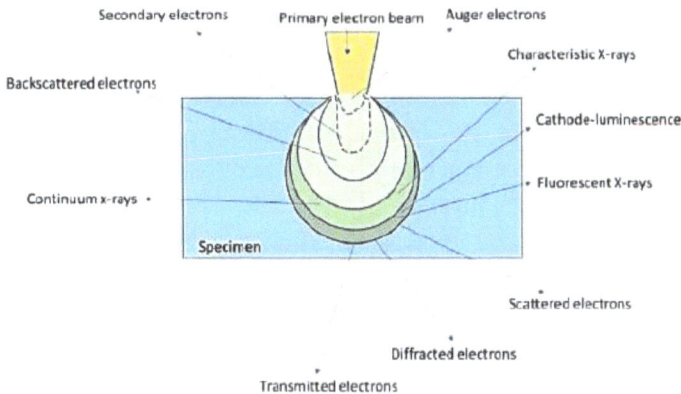

Fig. (6). Schematic representation of electron–matter interaction volume and types of the signal generated.

the various range of signals emitted by the excited atoms with the interaction of electron beam in a bulk specimen. Due to the interaction of the electron beam with the material, various signals are generated due to the excited atoms. The typical signals generated during the interaction are explained earlier. Out of various signals, secondary electrons are detected, and the imaging of the sample is obtained in the scanning electron microscope. The secondary electrons have very low energy in the order of 50 eV, which also limits the mean free path in the solid matter.

In general, the scanning electron microscopy (SEM) samples are prepared to have good electrical conductivity and withstand the vacuum conditions and high energy beam interactions. A typical SEM consists of the following components, as shown in Fig. (**7**). The electron beam is emitted from the electron gun and is guided by the condenser lens and aperture. The deflection coils are used for deflecting the beam to raster over the surface of the specimen and get the image.

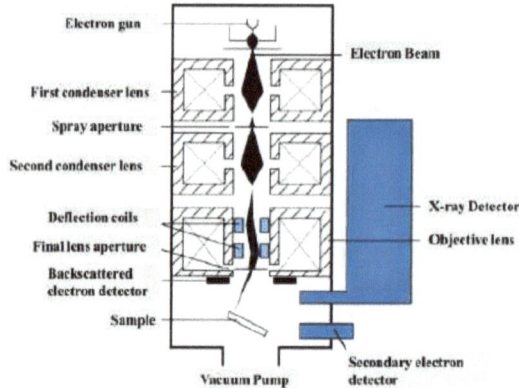

Fig. (7). Schematic representation of the scanning electron microscope.

SEM reveals the topographical details of the surface with many details with a high depth of focus compared to the optical microscope. In general, and analytical scanning electron microscopy gives the (a) secondary electron images of the surface features, (b) backscattered electron images of phase differentiation, precipitates, reaction regions based on average atomic number contrast, (c) topography images of pits, protrusions, reacted regions, *etc.*, (d) elemental distribution in near-surface regions with an additional attachment of EDS. The typical imaging modes in the SEM are discussed in Table **2**. Fig. (**8**) shows the depth of focus, which is large with a small aperture, and the focus is small with a large aperture size. However, in the case of an optical microscope, it depends on the aperture angle; if the angle is small, the depth of focus is large, and if the angle is large, the depth of the focus is small. The depth of focus can also be changed with magnification. Fig. (**9**) shows the difference in the depth of focus obtained by the optical microscope and the scanning electron microscope. The depth of focus is much higher in SEM compared to the optical microscope as the aperture angle of the electron probe in the SEM is much smaller than that of the optical lens [5]. The typical electron guns used in the SEM are shown in Table **2**.

Table 2. SEM imaging modes.

Mode	Information	Typical resolution	High Resolution
Backscattered electrons	Topographic, crystallographic and composition	10 nm	3 nm

(Table 2) cont.....

Mode	Information	Typical resolution	High Resolution
Secondary electrons	Topographic Voltage Magnetic and electric field	10 nm 100 nm 500 nm	3 nm 50 nm 100 nm
Absorbed specimen current	Topographic Composition	50 nm	20 nm

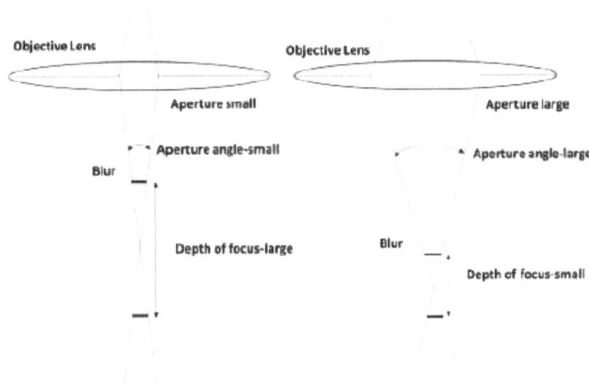

Fig. (8). Relationship between the aperture angle and the electron probe and the depth of focus.

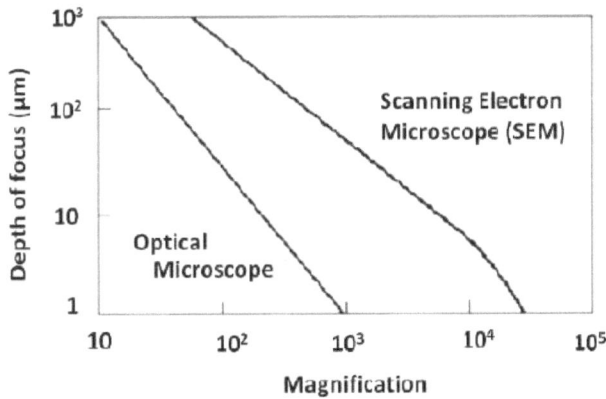

Fig. (9). Difference in the depth of focus in optical microscope and SEM.

Transmission Electron Microscopy

In the transmission electron microscope, the specimens are prepared as thin as 100 nanometers [6]. The interaction of the electron beam with the thin specimens eventually ends up heating the sample (photon excitation of the atomic lattice). However, before the electron comes to rest, it undergoes two types of scattering, elastic and inelastic, as shown in Fig. (**10**). Most of the techniques used to study the microstructure involve the use of certain forms of electromagnetic radiation and interaction with the specimen of different kinds. The various radiations used for microscopy and their mode of observation are presented in Table **3**. The inverse of the square of beam energy determines the elastic scattering cross cross-section. Therefore, the rate of elastic scattering per unit path length rapidly decreases as the beam energy increases. This reduction in elastic energy cross-section with increasing beam energy is being utilized in the thin foil transmission electron microscopy [6, 7]. There are various imaging modes in the TEM, as shown in Fig. (**11**) and typical TEM is shown in Fig. (**12**).

Table 3. Various Signals and mode of observation.

Nature of radiation		Mode of Observation
Particles	Wavelength	
X-ray photos	1	Diffraction
Electrons (high energy)	0.05	Microscopy and diffraction
Electrons (low energy)	1	Diffraction
Neutrons	1-10	Diffraction
Ions	-	Microscopy

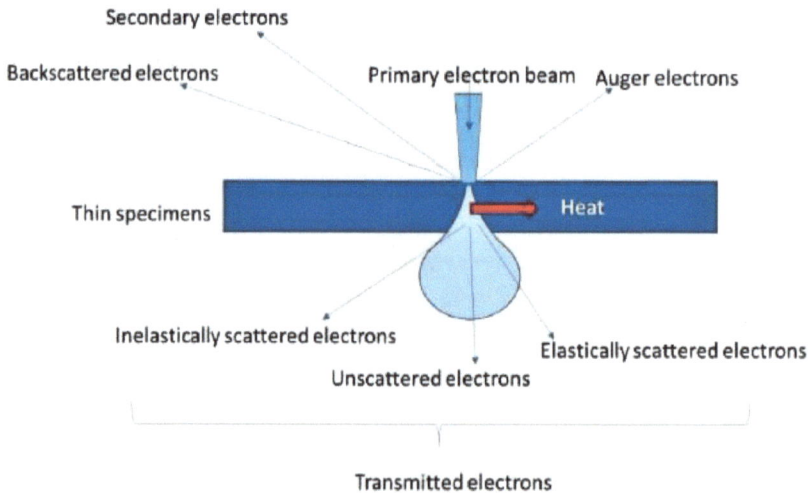

Fig. (10). Scattering of the transmitted electrons in the thin specimens.

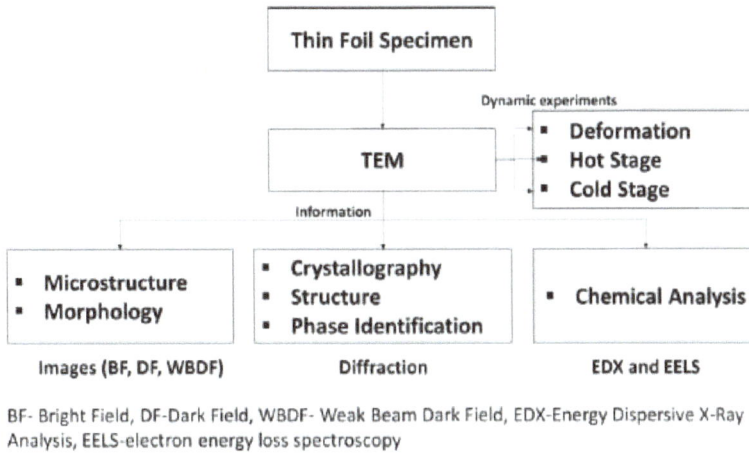

Fig. (11). Imaging modes and information extracted in the TEM.

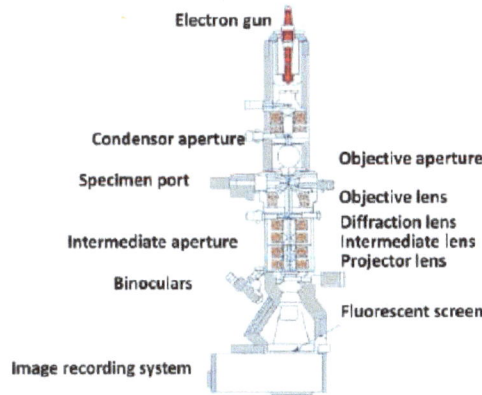

Fig. (12). Schematic representation of the transmission electron microscope.

TRIBOLOGICAL CHARACTERIZATION

Introduction

The word "Tribology", coined by Peter H Jost in the 1960s, derived from the Greek word, means science and technology of rubbing. This basically deals with friction, wear and lubrication. In tribology, studying the science of surfaces in contact is to understand the characteristics and sensitivity of the surfaces compared to bulk specimens. Depending on the application accuracy requirements, the surface has to be understood well from micro to nano-scale and has to be tailored accordingly. In engineering components like a cylinder - bore in Internal Combustion Engines, the hull of ships, optical components, *etc.*, the surface topography plays a key role [8 - 11, 19, 20].

The basic understanding of tribological aspects starts with the nature of the interaction of tribo-pair surfaces. Fig. **(13 (a), (b), and (c))** shows the schematic representation of point, line, and area type of contact of two body surfaces represented as 1 and 2, respectively, which are noticed in engineering applications. This classification of contact types was given when viewed on a macroscopic scale. But when the surfaces are observed on micro and nano scales, the asperities present on the surface get into contact. This summation of asperity level contacts is known as real contact area or true contact, whereas the area contacts on the bulk scale (as indicated in Fig. **(13 (c))**) can be referred to as an apparent contact area [8, 9]. The bodies in contact, as shown in Fig. **(13)**, are subjected to different types of loads in working conditions. During the static (load w.r.t time is constant) and dynamic (load w.r.t time is not constant; Fig. **(14)** interaction, depending on the stress intensity levels acting at contact interface and also elastic and plastic properties of the material pair, these asperities rub against each other, leading to friction and wear in dry condition [9, 12].

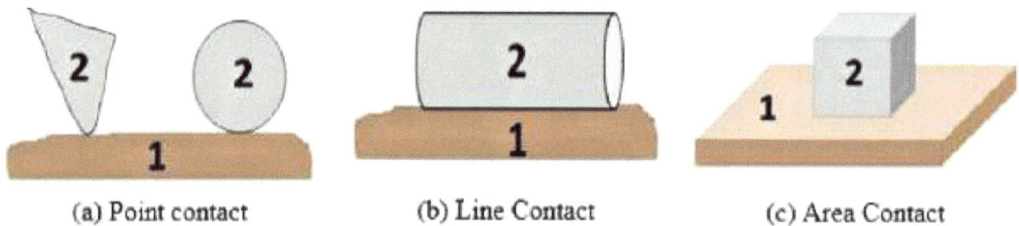

(a) Point contact   (b) Line Contact   (c) Area Contact

Fig. (13). Types of macro-scale contacts generally seen in tribological contacts (1 - First body & 2 - Second body).

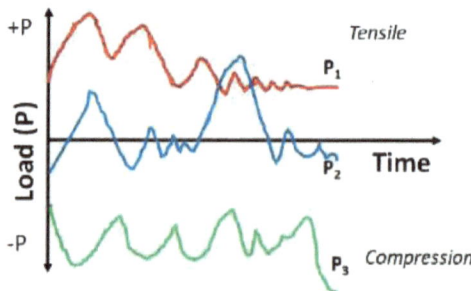

Fig. (14). Type of dynamic loads (P1, P2 and P3) generally seen in tribological contacts.

Classification of Tribology

The tribology is broadly classified into science and technology involved in the process, as shown in Fig. **(15)**. When some materials like quartz, *etc.*, are mechanically rubbed against each other, the surfaces get charged, electrons emission will happen, and optical light is generated. The physics involved in this process is tribo-physics and the corresponding sub-classifications are triboluminescence, and tribo-emission of electrons [15]. The chemistry part of tribology is tribo-chemistry, which deals with the chemical reactions as a result of energy, adsorption, desorption, diffusion and catalysis occurring at the interface zones of tribo-pair. These reactions cause physico-chemical changes along with mechanical changes on the layer of the tribo-pair surfaces, thereby influencing the friction and wear characteristics [16 - 18]. Under the technological aspects of tribology, the friction, wear and lubrication concepts are being studied intensively, and their individual sub-classification is shown in Fig. (**15**). Plenty of literature is available on the various types of friction, wear and lubrication mechanisms, including their sub-classification and detailed description of mechanisms, which can be found in some studies [8 - 14]. However, in this chapter, a few of them are reported briefly.

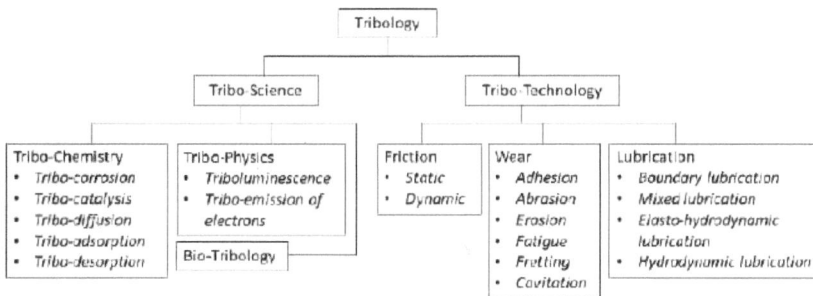

Fig. (15). Broad classification of tribology.

Surface Roughness Evaluation

In contact mechanics, surface structure plays an important role in understanding the mechanical behavior exhibited in between two solid objects interface, *i.e.*, non-contact to full contact of asperities. Surface roughness is the deviation of a real surface from its ideal form. It plays a significant role in determining how a real object will interact with its environment. It can be related closely to the friction and wear properties of a given surface. Rough surfaces usually have higher coefficients of friction and wear than smooth surfaces. Hence, controlled roughness is essential in some of the engineering applications like the engine cylinder bore of an internal combustion engine, bearings, *etc.* Nowadays, the surface roughness of a given sample can be evaluated by surface profilometer equipment of either contact, *i.e.*, physical movement of the probe across the surface or non-contact type, *i.e.*, by using optical light interference method. The quantification of surface roughness can be a single point, a line scan, or a full three-dimensional scan. The amplitude and frequency of the peaks and valleys present over the surface are evaluated. This is to get a complete description, *i.e.*, qualitative and quantitative information about the surface profile. The understanding of surface roughness will form a basis for studying friction and wear aspects of materials [8, 9].

Contact Type Surface Profilometer

In contact-type surface profilometer, a diamond stylus in contact with a sample surface is made to move physically in XYZ direction for a specified distance, contact force (up to 50 micrograms), and traverse rate to measure small vertical features over the surface. Fig. (**16**) shows the schematic and photographic image of the contact type surface profilometer. The size and shape of the stylus tip can influence the surface roughness measurements. This method can be destructive and not suitable for some surfaces, which are soft or visco-elastic in nature.

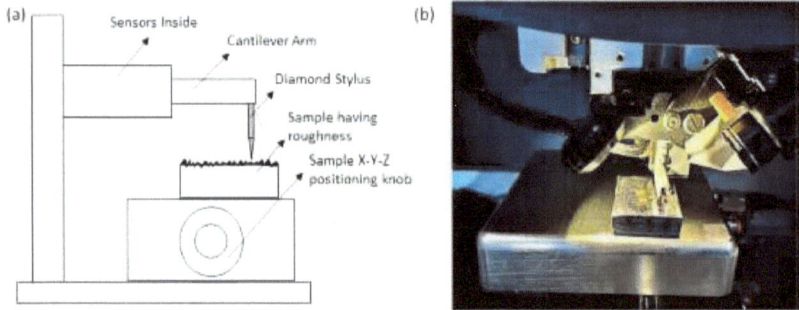

Fig. (16). (a) Schematic and **(b)** Photographic image of the tip in contact with the sample surface for surface roughness evaluation.

Non-Contact Type Surface Profilometer

In this type of surface profilometer, a light interference technique is used with the help of optical mirrors/lens. These profilometers do not touch the sample surface; hence the problem of sample damage can be avoided. The conventional instrument for measuring surface profile lies in its inability to evaluate samples of non-conventional type. The difficulty may arise due to changes in sample size, shape, damage of surface due to tip movement, and also sometimes part portability. To overcome these surface measurement problems, non-contact type surface profilometer is being used in industrial applications.

The exit of a light beam from the source is split into two beams by a beam separator, one to the reference and the other to the sample. The two beams after reflection from the sample and reference mirror are recombined for interference. This causes the formation of interference fringes, which are recorded by using a camera. The image of the sample surface is produced and analyzed. The various roughness parameters in use are Ra - Arithmetic average, Rq - Root mean squared, Rsk - Skewness, Rku - Kurtosis, Rv - Maximum valley depth, Rp -

Maximum peak height, Rc - Mean height, Rt - Total height, Rz - Maximum peak to valley height, Rmr - material ratio, Rdc - Profile section height difference. These parameters define the vertical (*i.e.*, amplitude) and horizontal characteristics (*i.e.*, spacing/material ratio) of a given surface profile. Fig. (17) shows the two-dimensional image of the surface and the corresponding roughness values of a metallurgically polished metal sample obtained from non - contact profilometer [8, 9, 14].

Friction and Wear Evaluation

Tribometer for Measuring Friction and Wear

The friction can be evaluated from a simple macro to nano level scratch testing under constant or progressive, or incremental loads. However, in order to study the wear aspects along with friction, a tribometer is needed. A tribometer is a piece of equipment that precisely measures the quantities like the coefficient of friction, friction forces, wear rate/volume/wear depth, and also other quantities like temperature and acoustic emission of a given bulk sample. The friction and wear depend on various parameters related to machine test conditions, relative motion, material aspects of the sample, geometry, type of contact, in dry contact or lubrication mode and environmental conditions existing between the tribo-pair and tribo - system. A lot of research has been done stating that the friction and wear will be affected due to relative motion, *i.e.*, sliding and/or rolling and/or reciprocating type, conformal or non-conformal contacts of tribo-pair and various test conditions like applied load, velocity, sliding distance, surface roughness, *etc* [8, 9].

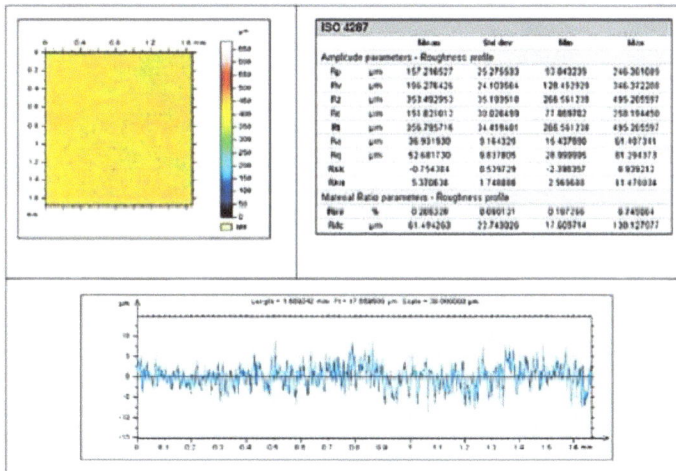

Fig. (17). An example of surface roughness data of the metallurgically polished sample.

Pin on disc tribometer consists of a stationary pin held normally against a rotating disc. The axis of the sample can be vertical or horizontal to the ground. This test equipment is for measuring sliding friction and wear. The severe forms associated with sliding wear are scuffing, galling, and scoring depending on the pressures applied and relative velocities between the tribo - pair. If a third body /hard particle is introduced at the interface of two surfaces, then the wear mechanism becomes abrasive wear. In this wear, the size, shape and hardness of the particles also define the severity of wear [10, 12]. Fig. (**18**) shows the SEM images of worn surface topology and wear debris obtained post wear testing. This will help in understanding the damage mechanisms and characteristics of sample surfaces. However, these performance characteristics differ when the testing of samples is done at elevated temperatures, vacuum/inert environments and also under lubrication mode [9, 14].

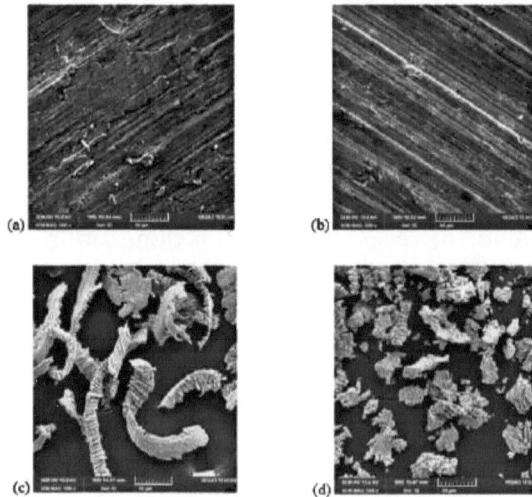

Fig. (18). (a)-(b) worn surface topology and **(c)-(d)** wear debris obtained post wear testing.

NANO-INDENTATION

In tribology, the mechanical state of the surface is very much important, and one has to understand well to know elastic limits, resistance to plastic deformation, brittleness, cracking, *etc.*, of a given material sample. This can be characterized by using four quantities, namely hardness, Young's Modulus, toughness, and residual stresses. In a hardness test, an indenter of a particular size and shape (Fig. **19**)) creates a residual impression on a material sample under a normal load to measure the surface area. The Young's Modulus along with ductility is measured from the stress-strain curves obtained through the uniaxial tensile test. The brittleness, which is the opposite of ductility, refers to the ease with which fractures propagate. The fracture propagation can be quantified through toughness parameter. Lastly, the internal residual stresses arise on the surface as a result of the production processes involved. These stresses facilitate or suppress the crack propagation and even cause a breakup of materials. The residual stresses can be measured by using X-ray diffraction technique.

Nanoindentation testing is done to evaluate the properties of the material for depths of nanometer to micrometer range. The main purpose of this test is to determine the hardness (H) and elastic modulus (E) of the given sample material from the measurements from load-displacement curves. The H of the specimen is determined by dividing the load of the indenter by the projected area of contact. The contact area size (A) can be determined from the penetration depth (hc) from the known indenter geometry. The E of the specimen can be resolved from the slope of the unloading curve of the P-h plot, as shown in Fig. (**19 (a)**) [14, 21 - 23]. The input parameters are the type of indenter tip (*i.e.*, Berkovich, Vickers pyramidal, spherical) and its material (Diamond, Si_3N_4), and load function (loading, dwell and unloading cycle). The output parameters are hardness, reduced modulus or elastic modulus, penetration depth, stiffness, elastic and plastic work. The factors influencing test data are surface roughness, indenter geometry (*i.e.*, size, shape and sharpness of tip), indentation size, residual stresses, surface forces, friction, and adhesion properties, piling and sinking effect on the material surface and drift behavior *i.e.*,, creep effects [21 - 23].

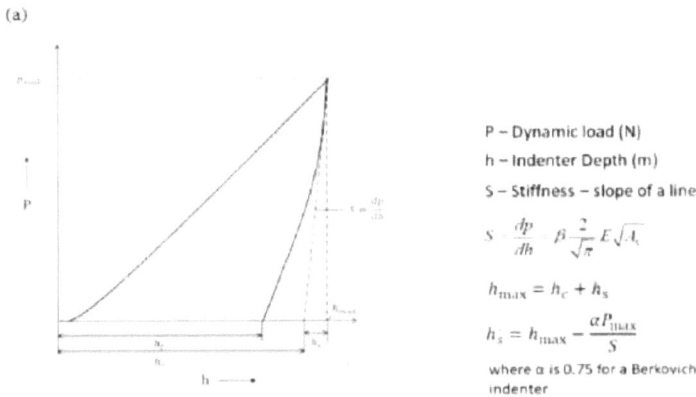

P – Dynamic load (N)
h – Indenter Depth (m)
S – Stiffness – slope of a line

$$S = \frac{dp}{dh} \quad \beta \frac{2}{\sqrt{\pi}} E \sqrt{A_c}$$

$$h_{max} = h_c + h_s$$

$$h_s = h_{max} - \frac{\alpha P_{max}}{S}$$

where α is 0.75 for a Berkovich indenter

Fig. (19). (a) P - h curve in Nanoindentation experiment **(b)** Schematic representation of surface deformed after the test.

ATOMIC FORCE MICROSCOPY

For studying nano mechanics and nanotribological properties on a given specimen, the widely used equipment are surface force apparatus (SFA), scanning tunneling microscopes (STM), and atomic force microscopes (AFM). Due to the sample limitations present in SFA and STM, like atomically smooth surfaces and electrically conductive surfaces, respectively, AFM is preferred. Fig. (**20**) shows the schematic of AFM indicating four quadrant photodetector, cantilever, AFM tip and XYZ piezo-electric tube [8, 24 - 26].

AFM can be used for research studies, including friction & wear, indentation, material transfer detection studies, fabrication purpose, *etc.*, with high resolution. The four-quadrant photodetector (Q1 to Q4) is used for detecting the vertical and lateral deviations of the reflected light beam. The vertical deviations [(Q1+Q2) - (Q3+Q4)] arise from the topological state of the surface, whereas lateral deviations [(Q1+Q3) - (Q2+Q4)] on the surface are used for friction measurements. As indicated in Fig. (**20**), an AFM probe consists of a cantilever protruded from a holder and a sharp tip on the free end of a cantilever. The schematic representation of different kinds of AFM tips is given in Fig. (**21**). Depending on application requirements, the AFM tips are being used for the study. Due to the interaction between the surface and the tip, the cantilever gets deflected and, in turn, signal deflection [8, 24 - 26].

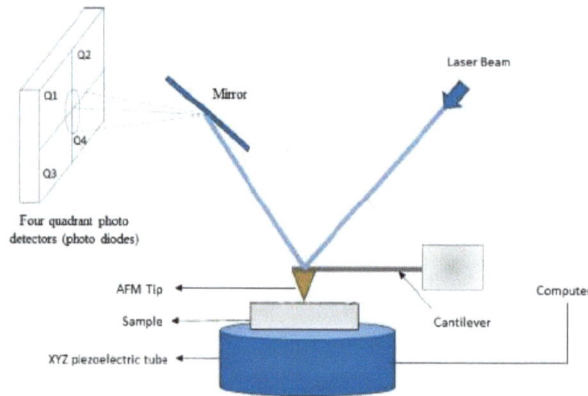

Fig. (20). Schematic representation of Atomic Force Microscopy (AFM).

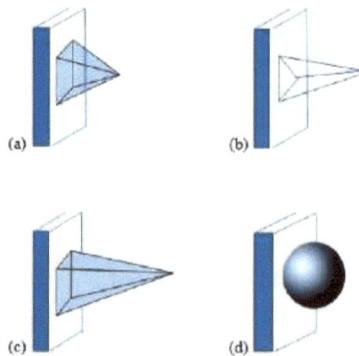

Fig. (21). Schematic representation of different types of AFM tips; (**a**) Square pyramidal tips (Silicon nitride), (**b**) Three-sided pyramidal diamond tip, (**c**) Square pyramidal (Single crystal silicon), (**d**) Spherical Colloidal probe.

Contact Mode

AFM tip will be in contact with the sample surface and dragged over the surface while in contact. The cantilevers of low stiffness for measuring enough large deflections are used. This is used for topography scan, lateral force or friction measurement, detection and mapping of electrostatic, magnetic, and also surface forces at the sample surface. Imaging at the atomic scale, measuring interfacial forces including friction and adhesion and surface roughness are the possible outcomes of this testing mode. The drawback of this mode is that it is not suitable for soft materials or soft surfaces, as they get scratched or deformed during tip travel [8, 24 - 26]. It is one of the characterization methods along with the SEM, EDS, XRD and others, which are widely used to characterize the materials and welded structures to find their structural properties [27 - 51].

Tapping Mode

This mode is also called dynamic contact mode and intermittent contact mode. In this mode, when the cantilever is at or near its resonance frequency, it is made to oscillate up and down, as shown in Fig. (**22**). This oscillation will be in the nanometer range. This mode is generally used for topography studies of soft surfaces, localized surface elasticity and viscoelastic mapping. The influence of the velocity of the tip movement on the sample is shown in Fig. (**23**). At low velocities, the tip trajectory will be influenced by atomic scale stick-slip (Fig. (**23(a)**)) if the surface is hydrophobic in nature or has hydrophobic film. This will lead to higher lateral force measurements. At higher velocities and if damping is low, the tip trajectory will experience tip jump of several cycles periodically. As a result, lower lateral force and lower friction measurement is observed (Fig. (**23(b)**)). The major abilities of an AFM are topology measurement and imaging, forces measurement, manipulation at atomic levels and stimulation of cells at a local level. AFM is being used to study the problems related to various disciplines of materials science, surface chemistry, physics, semiconductor materials, biology *etc* [8, 24 - 26].

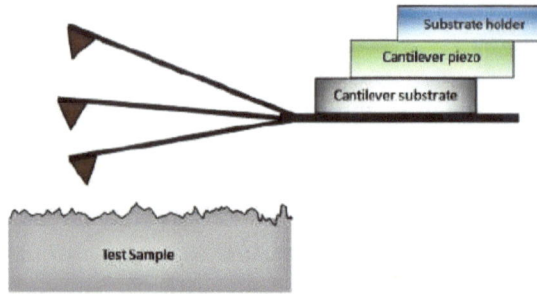

Fig. (22). Tapping mode operation in AFM.

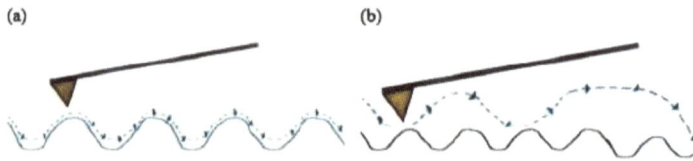

Fig. (23). (a) Tip Trajectory at low velocity (stick slip at the atomic scale), **(b)** Tip Trajectory at high velocity (a jump of the tip).

CONCLUDING REMARKS

In this chapter, the first part summarizes the history, principles, and applications of optical and electron microscopy. The optical microscope magnifies the object with the set of arrangements of the lens, however, it has restrictions in the resolution. This led to developments in the electron microscope, and the resolution range has been increased to study the interatomic spacing in the metals. In the bulk samples, up to certain centimeters in size can be studied in the scanning electron microscope and in the thin samples, the transmitted electrons can be captured, and high-resolution images can be obtained from the transmission electron microscopes. Depending on the application requirement, the kind of analysis on the specimens and resolution of various variants in the optical and electron microscopy are developed. In the second part of the chapter, tribological chararacterisation of surfaces was reported. To understand the friction and wear of surfaces, the hardness (*i.e.*, macro, micro and nano scale) and surface roughnesshave to be understood and quantified precisely. Hardness at the nano scale can be measured by using nanoindentation technique. Three-dimensional quantification of surface roughness can be done by either contact or non-contact surface profilometer. For topography scan, lateral/friction force measurements, and contact mode is used in AFM. For viscoelastic materials, tapping mode in AFM is used for topology measurement.

CONSENT FOR PUBLICATION

Not applicable.

CONFLICT OF INTEREST

The author declares no conflict of interest, financial or otherwise.

ACKNOWLEDGEMENTS

The authors would like to thank the management of BML Munjal University, Gurgaon, India, for their valuable support.

REFERENCES

[1]　Stevens K, Ed. Surface engineering. Springer Netherlands 1990.

[2]　Amelinckx S, Van Dyck D, Van Landuyt J, Van Tendeloo G, Eds. Handbook of Microscopy: Applications in Materials Science, Solid-State Physics, and Chemistry, Methods II. John Wiley & Sons 2008.

[3]　Török P, Kao FJ, Eds. Optical imaging and microscopy: techniques and advanced systems. Springer 2007.

[4]　Pennycook SJ, Nellist PD, Eds. Scanning transmission electron microscopy: imaging and analysis. Springer Science & Business Media 2011.
[http://dx.doi.org/10.1007/978-1-4419-7200-2]

[5]　Kushmerick JG, Weiss PS. Scanning probe microscopes. 2010: 2464-2472.

[6]　CARTER BA, Williams DB. Transmission Electron Microscopy: A Textbook for Materials Science. Diffraction II. Springer Science & Business Media 1996.

[7]　Brown PD. Transmission Electron Microscopy-A Textbook for Materials Science, by David B. Williams and C. Barry Carter. Microsc Microanal 1999; 5(6): 452-3.
[http://dx.doi.org/10.1017/S1431927699990529]

[8]　Bhushan B. Introduction to tribology. John Wiley & Sons 2013.
[http://dx.doi.org/10.1002/9781118403259]

[9]　Menezes PL, Nosonovsky M, Ingole SP, Kailas SV, Lovell MR, Eds. Tribology for scientists and engineers. New York: Springer 2013.
[http://dx.doi.org/10.1007/978-1-4614-1945-7]

[10]　Stachowiak G, Batchelor AW. Engineering tribology. Butterworth-Heinemann 2013.

[11]　Neale MJ. The tribology handbook. Elsevier 1995.

[12]　Hutchings I, Shipway P. Tribology: friction and wear of engineering materials. Butterworth-Heinemann 2017.

[13]　Davim JP, Ed. Tribology in manufacturing technology. Springer Science & Business Media 2012.

[14]　Takadoum J. Materials and surface engineering in tribology. John Wiley & Sons 2013.

[15]　Wu Junhui. Study on optimization of 3D printing parameters, IOP Conf. Series: Materials Science and Engineering 392
[http://dx.doi.org/10.1088/1757-899X/392/6/062050]

[16]　Kajdas C, Hiratsuka KI, Eds. Tribocatalysis, Tribochemistry, and Tribocorrosion. CRC Press 2018.
[http://dx.doi.org/10.1201/b20123]

[17]   Mang T, Bobzin K, Bartels T. Industrial tribology: Tribosystems, friction, wear and surface engineering, lubrication. John Wiley & Sons; 2011 Jan 19.

[18]   Jones MH, Scott D, Eds. Industrial tribology: the practical aspects of friction, lubrication and wear. Elsevier; 1983 Mar 1.

[19]   Roy M, Ed. Surface engineering for enhanced performance against wear. Heidelberg, Germany: Springer; 2013 Apr 4.
[http://dx.doi.org/10.1007/978-3-7091-0101-8]

[20]   Czanderna AW, Madey TE, Powell CJ, Eds. Methods of surface characterisation. Springer Science & Business Media; 2002 Apr 11.

[21]   Fischer-Cripps AC. Contact mechanics InNanoindentation. New York, NY: Springer 2011; pp. 1-19.

[22]   Oyen ML, Ed. Handbook of nanoindentation: with biological applications. CRC Press 2019.
[http://dx.doi.org/10.1201/9780429111556]

[23]   Bhushan B, Ed. Nanotribology and nanomechanics: an introduction. Springer 2017.
[http://dx.doi.org/10.1007/978-3-319-51433-8]

[24]   Kaupp G. Atomic force microscopy, scanning nearfield optical microscopy and nanoscratching: application to rough and natural surfaces. Springer Science & Business Media 2006.

[25]   Voigtländer B. Forces Between Tip and Sample InAtomic Force Microscopy. Cham: Springer 2019; pp. 161-76.
[http://dx.doi.org/10.1007/978-3-030-13654-3_10]

[26]   Santos NC, Carvalho FA. Atomic Force Microscopy. Berlin, Germany: Springer 2019.
[http://dx.doi.org/10.1007/978-1-4939-8894-5]

[27]   Cheepu M, Venkateswarlu D, Rao PN, Kumaran SS, Srinivasan N. Effect of process parameters and heat input on weld bead geometry of laser welded titanium Ti-6Al-4V alloy. InMaterials Science Forum 2019, 969, pp. 613-618. Trans Tech Publications Ltd.

[28]   Cheepu M, Venkateswarlu D, Rao PN, Kumaran SS, Srinivasan N. Optimization of process parameters using surface response methodology for laser welding of titanium alloy InMaterials Science Forum. Trans Tech Publications Ltd. 2019; Vol. 969: pp. 539-45.

[29]   Cheepu M, Srinivas B, Abhishek N, *et al.* Dissimilar joining of stainless steel and 5083 aluminum alloy sheets by gas tungsten arc welding-brazing process. InIOP Conference Series: Materials Science and Engineering 2018 Mar 1, 330(1), p. 012048. IOP Publishing.
[http://dx.doi.org/10.1088/1757-899X/330/1/012048]

[30]   Cheepu M, Che WS. Friction welding of titanium to stainless steel using Al interlayer. Trans Indian Inst Met 2019; 72(6): 1563-8.
[http://dx.doi.org/10.1007/s12666-019-01655-7]

[31]   Cheepu M, Venkateswarlu D, Rao PN, Muthupandi V, Sivaprasad K, Che WS. Microstructure characterization of superalloy 718 during dissimilar rotary friction welding InMaterials Science Forum. Trans Tech Publications Ltd. 2019; Vol. 969: pp. 211-7.

[32]   Anuradha M, Das VC, Susila P, Cheepu M, Venkateswarlu D. Microstructure and Mechanical Properties for the Dissimilar Joining of Inconel 718 Alloy to High-Strength Steel by TIG Welding. Trans Indian Inst Met 2020; 11: 1-5.
[http://dx.doi.org/10.1007/s12666-020-01925-9]

[33]   Anuradha M, Das VC, Susila P, Cheepu M, Venkateswarlu D. Effect of welding parameters on TIG welding of Inconel 718 to AISI 4140 steel. Trans Indian Inst Met 2020; 11: 1-6.
[http://dx.doi.org/10.1007/s12666-020-01926-8]

[34]   Muralimohan CH, Muthupandi V, Sivaprasad K. Properties of friction welding titanium-stainless steel joints with a nickel interlayer. Procedia Materials Science 2014; 5: 1120-9.

[http://dx.doi.org/10.1016/j.mspro.2014.07.406]

[35]   Muralimohan CH, Ashfaq M, Ashiri R, Muthupandi V, Sivaprasad K. Analysis and characterization of the role of Ni interlayer in the friction welding of titanium and 304 austenitic stainless steel. Metall Mater Trans, A Phys Metall Mater Sci 2016; 47(1): 347-59.
[http://dx.doi.org/10.1007/s11661-015-3210-z]

[36]   Muralimohan CH, Haribabu S, Reddy YH, Muthupandi V, Sivaprasad K. Evaluation of microstructures and mechanical properties of dissimilar materials by friction welding. Procedia Materials Science 2014; 5: 1107-13.
[http://dx.doi.org/10.1016/j.mspro.2014.07.404]

[37]   Cheepu M, Muthupandi V, Loganathan S. Friction welding of titanium to 304 stainless steel with electroplated nickel interlayer InMaterials Science Forum. Trans Tech Publications Ltd. 2012; Vol. 710: pp. 620-5.

[38]   Cheepu M, Ashfaq M, Muthupandi V. A new approach for using interlayer and analysis of the friction welding of titanium to stainless steel. Trans Indian Inst Met 2017; 70(10): 2591-600.
[http://dx.doi.org/10.1007/s12666-017-1114-x]

[39]   Muralimohan CH, Muthupandi V, Sivaprasad K. The influence of aluminium intermediate layer in dissimilar friction welds. International Conference on Engineering Materials and Processes. 350-7.

[40]   Lee KH, Choi SW, Yoon TJ, Kang CY. Microstructure and hardness of surface melting hardened zone of mold steel, SM45C using Yb: YAG disk laser. Journal of Welding and Joining 2016; 34(1): 75-81.
[http://dx.doi.org/10.5781/JWJ.2016.34.1.75]

[41]   Choi D, Shin J. Weld Shape Analysis using Central Composite Design in the Laser Welding of Aluminum Alloys. Journal of Welding and Joining 2020; 38(5): 502-7.
[http://dx.doi.org/10.5781/JWJ.2020.38.5.10]

[42]   Yun T-J, Oh W-B, Lee B-R, *et al.* A Study on Optimization of Fillet in Laser Welding Process for 9% Ni Steel Using Gradient-Based Optimization Algorithm. Journal of Welding and Joining 2020; 38(5): 485-92.
[http://dx.doi.org/10.5781/JWJ.2020.38.5.8]

[43]   Kim J-Y, Lee D-M. LAM-DED Process for Repair and Maintenance of Cast Iron Components using Metallic Powder Alloys. Journal of Welding and Joining 2020; 38(4): 349-58.
[http://dx.doi.org/10.5781/JWJ.2020.38.4.3]

[44]   Cheepu M, Susila P. Interface microstructure characteristics of friction-welded joint of titanium to stainless steel with interlayer. Trans Indian Inst Met 2020; 7: 1-5.
[http://dx.doi.org/10.1007/s12666-020-01895-y]

[45]   Cheepu M, Che WS. Characterization of microstructure and interface reactions in friction welded bimetallic joints of titanium to 304 stainless steel using nickel interlayer. Trans Indian Inst Met 2019; 72(6): 1597-601.
[http://dx.doi.org/10.1007/s12666-019-01612-4]

[46]   Jeyaprakash N, Yang CH, Duraiselvam M, Sivasankaran S. Comparative study of laser melting and pre-placed Ni–20% Cr alloying over nodular iron surface. Arch Civ Mech Eng 2020; 20(1): 1-2.
[http://dx.doi.org/10.1007/s43452-020-00030-4]

[47]   Jeyaprakash N, Yang CH, Duraiselvam M, Prabu G, Tseng SP, Kumar DR. Investigation of high temperature wear performance on laser processed nodular iron using optimization technique. Results Phys 2019; 15: 102585.
[http://dx.doi.org/10.1016/j.rinp.2019.102585]

[48]   Cheepu M, Kumar Reddy YA, Indumathi S, Venkateswarlu D. Laser welding of dissimilar alloys between high tensile steel and Inconel alloy for high temperature applications. Advances in Materials and Processing Technologies 2020.

[49]   Cheepu M, Venkateswarlu D, Rao PN, Kumaran SS, Srinivasan N. Effect of process parameters and

heat input on weld bead geometry of laser welded titanium Ti-6Al-4V alloy. InMaterials Science Forum . 969: 613-8.

[50]   Chandra GR, Venukumar S, Cheepu M. Influence of rotational speed on the dissimilar friction welding of heat-treated aluminum alloys. InIOP Conference Series: Materials Science and Engineering 2020 Dec 1, 998(1), p. 012070. IOP Publishing.

[51]   Cheepu M, Cheepu H, Che WS. Influence of joint interface on mechanical properties in dissimilar friction welds. Advances in Materials and Processing Technologies 2020.

SUBJECT INDEX